CIRCLE OF POISON

CIRCLE OF POISON

Pesticides and People
in a Hungry World

By David Weir and
Mark Schapiro

INSTITUTE FOR FOOD AND DEVELOPMENT POLICY
1885 MISSION STREET, SAN FRANCISCO, CA 94103 U.S.A. (415) 864-8555

Library of Congress Cataloging in Publication Data

Weir, David, 1947-
  Circle of Poison

  Includes bibliographical references
  1. Pesticides--Toxicology. 2. Pesticides--Environ-
mental aspects. 3. Pesticides--Industry. I. Schapiro,
Mark, 1955-   II. Institute for Food and Development Policy (San Francisco, Calif.) III. Title.
RA1270.P4W4   363.1'7   81-13384
ISBN 0-935028-09-9 (pbk.)   AACR2

To order additional copies, write or call:

Institute for Food and Development Policy
2588 Mission Street
San Francisco, CA 94110 USA
(415) 648-6090

Please add 15 percent for postage and handling
($1 minimum). Bulk discounts available.

Cover and book design by Kerry Tremain
Production by Claire B. Greensfelder
Cover drawing by Kirk Ditzler
Cover photograph by Nick Allen

ACKNOWLEDGMENTS Our

thanks to everyone on the staffs of the Institute for Food and Development Policy and the Center for Investigative Reporting. At the Institute, especially Frances Moore Lappé for her editing, Joe Collins for his editorial ideas and information from overseas, David Kinley for his information from India, Doug Basinger, Jennifer Lovejoy and Stephen King for their copyediting and production help, and Gretta Goldenman for her work to publicize the book.

At the Center, Dan Noyes, Becky O'Malley and Maureen O'Neill Holcher, for their support during the long period we worked on this book.

In addition, we appreciate the editorial assistance of Nick Allen.

There were many, many others, both here and overseas, who supplied information or helped in other ways to make this book possible. They include Alberto Donadio, Daniel Samper and Elkin Bustamante (Colombia), Fernando Mazariegos and Jerry La Gra (Guatemala), W. F. Almeida (Brazil), Samuel Gitonga and Julie Kosterlitz (Kenya), Luise Custer, the Consumer Association of Penang (Malaysia), the Farmer's Assistance Board (Philippines), and many others whose names cannot appear here. We are especially indebted to those who supplied information but must remain anonymous for reasons of personal and political safety.

In the U.S. we thank Howard Bray of the Fund for Investigative Journalism, Benton Rhoades of the National Council of Churches, Terry Jacobs, Jacob Scherr, Noel Brown, Maureen Hinkle, Barry Castleman, Doug Shane, Ralph Lightstone, Ellen Widess, Mike Schneider, Skip Stiles, Mary Shinoff, Herb Mills, Raul Ramirez, Chris Jenkins, Fred Goff, Howard Kohn, Diana Kohn, Linda French, and Alison Weir, who first noticed that packet of Kool-Aid.

This project would not have been possible without the financial support of the Fund for Investigative Journalism, Agricultural Missions of the National Council of Churches, United Presbyterian Church in the U.S.A. and the Tides Foundation. **David Weir and Mark Schapiro.**

CONTENTS

PREFACE

Pat Goudvis

Box of DDT *being shipped from*
San Francisco.

T̲EN YEARS AGO, not long after the U.S. government banned cyclamates from food products as a potential cancer risk, David Weir, then an English teacher for the U.S. Peace Corps, bought a package of Kool-Aid in a small bazaar shop in Afghanistan. Reading the label, he was surprised to find cyclamates listed among the ingredients—somehow the package had found its way to the bazaars of Central Asia after the U.S. ban. Several years later, *Wall Street Journal* reporter Stan Sesser solved the mystery. He reported that after the ban, U.S. corporations had deliberately dumped products containing cyclamates on overseas markets.

That Kool-Aid packet launched Weir's decade-long series of inquiries, culminating in this book, co-authored with Mark Schapiro.

Weir began to investigate the problem of banned exports while he was an editor of two small San Francisco publications, *SunDance* and *Pacific Basin Reports*. Although his attempts at that time were constrained by a lack of resources, he did gather valuable files for later use. Then, while an editor of *Rolling Stone* in 1975–76, Weir had his first real breakthrough. Through a series of interviews in New York and Washington, he developed sources in South America who supplied evidence of banned drugs and pesticides sold by U.S. corporations there. His article, "For Export Only: Poisons and Dangerous Drugs," was published in early 1977. The story was not picked up by the U.S. press. But it was noticed in countries most affected by the problem: it was translated and reprinted in Mexico and Colombia. A four-page summary of the piece also appeared in the President's weekly magazine summary.

During the summer of 1977, Weir also wrote two articles for Pacific News Service about the contamination of imported coffee with residues of banned pesticides. These attracted considerable press interest, and coincided with the first attempt to close the loophole in U.S. legislation allowing banned pesticides to be exported—an attempt spearheaded by the Natural Resources Defense Council in Washington, D.C., especially attorney Jacob Scherr.

In 1978 the hazardous export issue exploded into the public arena. It was disclosed that Tris-treated baby clothes were being dumped on third world countries after the cancer-causing chemical Tris was banned by the U.S. government. Congress held hearings and issued a report about hazardous exports and President Jimmy Carter formed an inter-agency task force to assess the problem. The main impetus for Washington's interest was the fear that "Made in USA" time bombs might be ticking away throughout the third world, endangering this nation's foreign relations.

Accordingly, the State Department sponsored a conference in mid-1979 to examine the specific problem of hazardous pesticide exports to the third world. Scientists, corporate representatives, environmentalists, third world and U.S. government officials, and a small group of observers and reporters (including Weir) attended the conference, exchanging views on the subject.

In the meantime, *Mother Jones* magazine asked Weir, Mark Schapiro and their colleagues at the Center for

Investigative Reporting to provide basic research on the hazardous exports issue, as well as write a major article on pesticides in the third world. Co-authored with Terry Jacobs of the Center staff, this article appeared as part of a package on banned exports, entitled "The Corporate Crime of the Century" in the November 1979 issue of the magazine.

This package provoked an explosive reaction both here and overseas—a reaction which has not yet subsided. The United Nations passed a resolution on hazardous exports; new, tough legislation was introduced in Congress by Rep. Michael Barnes; the press heavily reported the stories; several third world governments issued new regulatory standards for pesticides, and activists here and abroad opened new initiatives to bring pressure on government agencies and the corporations responsible for the sales of dangerous products in the third world. Simultaneously, the Environmental Protection Agency enacted new regulations slightly tightening the loophole which allows banned pesticides to be exported.

The *Mother Jones* package won the 1980 National Magazine Award for reporting and was a finalist in the public service category. It was also named the "Best Censored" story of the year (an award the *Rolling Stone* story also won three years earlier) by a panel of judges selecting the most important stories ignored or underplayed by the press.

Early this year Frances Moore Lappé and Joseph Collins of the Institute for Food and Development Policy asked Weir and Schapiro to expand upon their earlier work to provide a tool for mobilizing public concern about the ongoing problem of pesticide dumping. By reinterviewing over a hundred sources, finding new reports, studies, people and documentation, Weir and Schapiro, with the editorial assistance of Frances Moore Lappé, Doug Basinger and Nick Allen, have produced this volume.

Here pesticides are the dish of the day, and one swallows more poison than food. There is not even a living hen or pig, and lately even the children are often sick. Could it be that even the gift that God gives—children—we cannot have?

ALFONSO CASTRO, COLOMBIAN FARMER

There's no problem with the ban of DBCP within the United States. In fact, it was the best thing that could have happened to us. You can't sell it here anymore but you can still sell it anywhere else. Our big market has always been exports anyway.

EXECUTIVE, AMVAC CORPORATION

Small shops in Indonesia sell pesticides right alongside the potatoes and rice. The people just collect it in sugar sacks, milk cartons, Coke bottles—whatever is at hand.

LUCAS BRADER, UN FOOD AND AGRICULTURE
ORGANIZATION

Nearly half of the green coffee beans imported into the United States contain various levels—from traces to illegal residues— of pesticides that have been banned in the United States.

FOOD AND DRUG ADMINISTRATION

CHAPTER ONE
THE CIRCLE OF POISON

This pesticide barrel in a mountain village in El Salvador is used to collect rainwater that is used for drinking by school children in the adjacent building.

THIS BOOK documents a scandal of global proportions—the export of banned pesticides from the industrial countries to the third world. Massive advertising campaigns by multinational pesticide corporations—Dow, Shell, Chevron—have turned the third world into not only a booming growth market for pesticides, but also a dumping ground. Dozens of pesticides too dangerous for unrestricted use in the United States are shipped to underdeveloped countries. There, lack of regulation, illiteracy, and repressive working conditions can turn even a "safe" pesticide into a deadly weapon.

According to the World Health Organization, someone in the underdeveloped countries is poisoned by pesticides *every minute.*[1]

But we are victims too. Pesticide exports create a circle of poison, disabling workers in American chemical plants and later returning to us in the food we import. Drinking a morning coffee or enjoying a luncheon salad, the American consumer is eating pesticides banned or restricted in the United States, but legally shipped to the third world. The United States is among the world's top food importers

and 10 percent of our imported food is officially rated as
contaminated.[2] Although the Food and Drug Administration (FDA) is supposed to protect us from such hazards,
during one 15-month period, the General Accounting Office (GAO) discovered that *half* of all the imported food
identified by the FDA as pesticide-contaminated was marketed without any warning to consumers or penalty to importers.[3]

*At least 25 percent of U.S. pesticide exports are products that
are banned, heavily restricted, or have never been registered for
use here.*[4] Many have not been independently evaluated for
their impacts on human health or the environment. Other
pesticides are familiar poisons, widely known to cause cancer, birth defects and genetic mutations. Yet, the Federal
Insecticide, Fungicide, and Rodenticide Act explicitly
states that banned or unregistered pesticides are legal for
export.[5]

In this book we concentrate on hazardous pesticides
which are either banned, heavily restricted in their use, or
under regulatory review in the United States. Some, such
as DDT, are banned for any use in the United States;
others, such as 2,4-D or toxaphene, are still widely used
here but only for certain usages. (Table I, on page 79, lists
the status of hazardous pesticides which are prohibited, restricted, under review, or unrestricted in the United
States.) As we will discuss, even "safe" pesticides which are
unrestricted in the United States may have much more
damaging effects on people and the environment when
used under more brutal conditions in the third world.

In the United States, a mere dozen multinational corporations dominate the $7-billion-a-year pesticide market.
Many are conglomerates with major sales in oil, petro-
chemicals, plastics, drugs and mining.

The list of companies selling hazardous pesticides to the
third world reads like a Who's Who of the $350-billion-
per-year[6] chemical industry: Dow, Shell, Stauffer, Chevron, Ciba-Geigy, Rohm & Haas, Hoechst, Bayer, Monsanto, ICI, Dupont, Hercules, Hooker, Velsicol, Allied,
Union Carbide, and many others. (See Table One.)

Tens of thousands of pounds of DBCP, heptachlor,

chlordane, BHC, lindane, 2,4,5-T and DDT are allowed to
be exported each year from the United States, even
though they are considered too dangerous for unrestricted
domestic use.[7]

"You need to point out to the world," Dr. Harold Hubbard of the U.N.'s Pan American Health Organization told
us, "that there is absolutely no control over the manufacture, the transportation, the storage, the record-keeping—
the entire distribution of this stuff. These very toxic pesticides are being thrown all over the world and there's no
control over any of it!"[8]

Not only do the chemical corporations manufacture
hazardous pesticides, but their subsidiaries in the third
world import and distribute them. (See Table One.)

• Ortho: In Costa Rica, Ortho is the main importer of
seven banned or heavily restricted U.S. pesticides—DDT,
aldrin, dieldrin, heptachlor, chlordane, endrin, and BHC.
Ortho is a division of Chevron Chemical Company, an arm
of Standard Oil of California.[9] (See Appendix B.)

• Shell, Velsicol, Bayer, American Cyanamid, Hercules
and Monsanto: In Ecuador these corporations are the main
importers of pesticides banned or restricted in the United
States—aldrin, dieldrin, endrin, heptachlor, kepone, and
mirex.[10]

• Bayer and Pfizer: In the Philippines, these multinationals import methyl parathion and malathion respec-
tively;[11] neither is banned but both are extremely hazardous.

The Ministry of Agriculture of Colombia registers 14
multinationals which import practically all the pesticides
banned by the United States since 1970.[12] (See Appendix
C.) And in the Philippines, the giant food conglomerate
Castle & Cooke (Dole brand) imports banned DBCP for
banana and pineapple operations there.[13]

Pesticides: a pound per person

WORLDWIDE pesticide sales are exploding. The
amount of pesticides exported from the U.S. has
almost doubled over the last 15 years.[14] The industry now

produces four billion pounds of pesticides each year—
more than one pound for every person on earth. [15] Almost
all are produced in the industrial countries, but 20 percent
are exported to the third world. [16]

And the percentage exported is likely to increase rap-
idly: The GAO predicts that during the decade ending in
1984, the use of pesticides in Africa, for example, will
more than quintuple. [17] As the U.S. pesticide market is
"approaching saturation . . . U.S. pesticide producers have
been directing their attention toward the export potential
. . . exports have almost doubled since 1965 and currently
account for 30 percent of total domestic pesticide production," the trade publication *Chemical Economics Newsletter*
noted. [18]

Corporate executives justify the pesticide explosion
with what sounds like a reasonable explanation: the hungry world needs our pesticides in its fight against famine.
But their words ring hollow: in third world fields most
pesticides are applied to luxury, export crops, not to food
staples the local people will eat. Instead of helping the poor
to eat better, technology is overexposing them to chemicals that cause cancer, birth defects, sterility and nerve
damage.

"Blind" schedules, not "as needed"

BUT THE CRISIS is not just the export of banned pesticides. A key problem in both the industrial countries
and the third world is the massive overuse of pesticides
resulting from their indiscriminate application. Pesticides
are routinely applied according to schedules preset by the
corporate sellers, not measured in precise response to actual pest threats in a specific field. By conservative estimate, U.S. farmers could cut insecticide use by 35 to 50
percent with no effect on crop production, simply by treat-
ing only when necessary rather than by schedule. [19] In Central America, researchers calculate that pesticide use, especially parathion, is 40 percent higher than necessary to
achieve optimal profits. [20]

In the United States the result of pesticide overuse is the

unnecessary poisoning of farmworkers and farmers— about 14,000 a year according to the Environmental Protection Agency (EPA).[21] But if pesticides are not used safely here—where most people can read warning labels, where a huge government agency (the EPA) oversees pesticide regulation, and where farmworker unions are fighting to protect the health of their members—can we expect these poisons to be used safely in the third world?

An inappropriate technology

IN THIRD WORLD countries one or two officials often carry responsibility equivalent to that of the entire U.S. EPA. Workers are seldom told how the pesticides could hurt them. Most cannot read. And even if they could, labels on banned pesticides often do not carry the warnings required in the United States. Frequently repacked or simply scooped out into old cans, deadly pesticides are often handled like harmless white powder by peasants who have little experience with manmade poisons.

But perhaps even more critical is this question: can pesticides—poisons, by definition—be used safely in societies where workers have no right to organize, no right to strike, no right to refuse to carry the pesticides into the fields? In the Philippines, for example, at least one plantation owner has reportedly sprayed pesticides on workers trying to organize a strike.[22] And, in Central America, says entomologist Lou Falcon, who has worked there for many years, "The people who work in the fields are treated like half-humans, slaves really. When an airplane flies over to spray, they can leave if they want to. But they won't be paid their seven cents a day or whatever. They often live in huts in the middle of the field, so their homes, their children, and their food all get contaminated."[23]

Yet the President's Hazardous Substances Export Policy Task Force predicts that the export of banned pesticides is likely to increase as manufacturers unload these products on countries hooked on the ag-chemical habit. "Continued new discoveries of carcinogenic and other damaging effects of many substances are probable over the next few years," predicts the task force. "In some cases, certain firms may be left with stocks of materials which can no longer be sold in the United States, and the incentive to recover some of their investment by selling the product abroad may be considerable." [24]

The genetic boomerang

T HE PESTICIDE EXPLOSION also has a second built-in boomerang. Besides the widespread contamination of imported food, the overuse of hazardous pesticides has created a global race of insect pests that are resistant to pesticides. The number of pesticide-resistant insect species doubled in just 12 years—from 182 in 1965 to 364 in 1977, according to the U.N. Food and Agriculture Organization.   So more and more pesticides—including new, more potent ones—are needed every year just to maintain present yields.

A circle of victims

B UT ENORMOUS DAMAGE is done even before the pesticides leave American shores. At Occidental's DBCP plant in Lathrop, California, workers discovered too late that they were handling a product which made them sterile. Elsewhere in the United States, worker exposure to the pesticides Kepone and Phosvel resulted in terrible physical and mental damage.

As part of our investigation into the "circle of poison," we looked at these examples of how the manufacture of hazardous pesticides affects American workers. Since companies are allowed to produce pesticides for export without providing health or safety data, there is no way to be sure they are not poisoning their own workers in the process. In fact there is abundant evidence that workers in the industrial countries are indeed suffering from their employers' booming export sales.

We talked with West Coast pesticide workers who complained of inadequate protection—and information—on the job. Even after two hot showers, one group explained, their hands still carried enough toxic residue of an unregistered pesticide that, when they stuck a finger in a fish bowl, the goldfish died.

These workers in pesticide manufacturing plants are the very first victims in the circle of poison. Add to them all the people who load and unload the chemicals into and out of trucks, trains, ships and airplanes; and those who have to clean up the toxic spills which inevitably occur. Then the total number of potential American victims of hazardous pesticide exports becomes very large. In addition, of course, there are the victims whose story is told in this book—third world peasants, workers and consumers—as well as everyone else in the world who eats food contaminated with pesticide residues. *We* complete the circle of victims.

To uncover the story in this book we have had to overcome powerful obstacles. The pesticide industry is a secretive one. The Environmental Protection Agency guards industry production data from the public, press and even other government agencies. The information made available often seems to defy meaningful interpretation.

We filed over 50 requests under the Freedom of Information Act in order to penetrate the industry's "trade secrets" sanctuary inside the EPA and other agencies. In assembling hundreds of tiny pieces of the puzzle, we studied trade publications and overseas magazines and newspapers for evidence of hazardous pesticide sales. In addition, we obtained import figures from a number of third world countries. Finally, we interviewed hundreds of people in industry, government, unions, environmental groups and international organizations. We corresponded with farmers, consumers and environmental groups in the third world.

The story told here is intended not merely to shock and to outrage. Its purpose is to mobilize concerned people everywhere to halt the needless suffering caused by pesticides' circle of poison.

CHAPTER TWO

A VICTIM
EVERY
MINUTE

Guatemala News & Information Bureau

Cotton workers near Esquintla, Guatemala.

FOR CHEMICAL COMPANY executives, exporting hazardous pesticides is not "dumping." If one country bans your product, move to where sales are still legal. It's just good business. But "good business" practice seldom takes account of the human toll inflicted by the massive use of pesticides.

As we mentioned in Chapter One, every minute of the day, on the average, someone is poisoned by pesticides in the third world. This World Health Organization statistic amounts to 500,000 poisoned people every year. A pesticide-caused death occurs about every hour and 45 minutes, totaling at least 5,000 each year.[1] Yet these estimates tell us nothing about the number of cancers, miscarriages, deformed babies and still-births resulting from the use of pesticides.

The rate of pesticide poisoning in underdeveloped countries is more than 13 times that in the United States, despite vastly greater use here, according to Virgil Freed, a consultant to the U.S. Agency for International Development (AID).[2] But *why* are there so many more victims in the third world? The following accounts from around the world tell why.

Culiacán, Mexico

IN CULIACAN in Northern Mexico, where large plantations grow tomatoes for American supermarkets, government doctors report seeing two or three pesticide poisonings every week. Sometimes workers are brought in
with convulsions. Since they get no paid sick leave, often
they return immediately to the fields, where their condition deteriorates. Every two or three weeks a federal hospital in Culiacán treats a farmworker for aplastic anemia, a
blood disease linked to organochlorine pesticides used in
the area. About half of these victims die.[3]

But *Los Angeles Times* reporters Laurie Becklund and
Ron Taylor were told by one group of workers that "someone in their camp dies every two or three days."[4] The farmworkers are routinely poisoned by drifting pesticide sprays
and leaking pesticide applicators, according to the reporters.

The workers live along the small patches of earth between the crops and the irrigation canals that receive all of
the pesticide run-off. "They wash their babies, their dishes
and their clothes in the canals and then turn back to the
canals to fill discarded insecticide tubs with canal water to
drink," reports the *Times*. While the workers become ill
from contaminated water, modern greenhouses with puri-
fied water systems have been erected to nurture the to-
mato seedlings. "The seedlings are more important than
the people," one U.S.-born grower explained.[5]

Central America

MORE THAN 14,000 poisonings and 40 deaths from
pesticides were tabulated between 1972 and 1975 in
the cotton-growing Pacific coastal plains of Central America, according to a 300-page report by the Central American Institute of Investigation and Industrial Technology
(ICAITI).[6] The actual total is undoubtedly much higher,
but impossible to determine. According to the report,
"some of the large cotton producers maintain their own
clinics [partly] to hinder public health officials from de-

tecting the seriousness of human insecticide poisonings."[7]

Although the pesticides are applied mainly to cotton grown for export, food crops—mainly corn and beans—are often contaminated simply because they are near the cotton fields. The report says that 75 percent of the sprayed pesticide frequently misses the cotton fields completely.[8] And toxic residues contaminate the soil.

Some farmworkers try to wash the pesticide from their skin, the ICAITI study revealed. But they use the irrigation drainage ditches, laced with the toxic runoff of insecticides, thereby compounding their contamination. Washing could not remove much of the parathion anyway, due to its pernicious tendency to concentrate in the oil on the skin, which transmits it directly into the bloodstream.[9]

Parathion, which causes 80 percent of Central America's poisonings,[10] was originally developed for chemical warfare by Nazi scientists during World War II. Slight chemical alterations converted it into a profitable insecticide after the war. The lethal dose of parathion to human beings is about *one-sixtieth* that of DDT: that is, it is 60 times more toxic. Parathion, explains Dr. H. L. Falk, of the National Institute of Environmental Sciences, "breaks down the substance which your body produces to stop the movement of your finger or your eye, for example. So those movements won't stop. You exhaust the muscles until they stop functioning altogether. You go into convulsions and die."[11]

The legacy of heavy pesticide use in Central America is ominous. Average DDT levels in cow's milk in Guatemala are 90 times as high as allowed in the United States. People in Nicaragua and Guatemala carry 31 times more DDT in their blood than people in the United States, where the substance has been banned since 1970.[12]

In Guatemala, reports *New York Times* correspondent Alan Riding, "the worst conditions, though the best pay, are on the cotton plantations. Here, pesticide spraying levels are so high that shipments of meat from cattle ranches in the area are frequently rejected by the United States Department of Agriculture because of their high DDT content. Studies also show that DDT levels in human blood in the cotton districts are eight times higher than in Guatemala City. Yields, though, are among the highest in the world. 'It's very simple,' explained Eduardo Ruiz, a young cotton planter. 'More insecticide means more cotton, fewer insects mean higher profits.'[13]

"But little concern is shown for those living and working in the region," reports Riding. "At the height of spraying [in the Tiquisate area], 30 or 40 people are treated daily in the nearby government clinic for the toxic effects on the liver and other organs.

" 'The farmers often tell the peasants to give another reason for their sickness, but you can smell the pesticide in their clothing,' a nurse said. 'And we know the symptoms— dizziness, vomiting and weakness. Only people who die in the clinic are reported. Otherwise bodies are buried on the farms.' "[14]

Pakistan

A WORLD AWAY from Central America, pesticides also kill. In Pakistan, at least five persons died and 2,900 others became ill in 1976 from malathion supplied in part by New York–based American Cyanamid for a U.S. government program to eradicate malaria.[15] Monte-Edison, an Italian chemical company, also supplied the malathion.

Government silence

F EW THIRD WORLD countries have either adequate pesticide regulations or the capacity to enforce them. As a result, the multinational pesticide producers have a free hand. Central America, for instance, has been turned into "a sort of experimental grounds for pesticide manufacturing companies," concludes the detailed study cited earlier.[16]

Most third world governments are reluctant to disclose their poisoning statistics, incomplete as they might be. Robert Chambers, who supervised the GAO's investigation of pesticides, cites three reasons the pesticide poisonings are often hushed up.

"One is tourism," he explains. "It doesn't look good to have press reports about contaminated food. Two, no government wants to admit it was poisoning its own people. Would you admit you were allowing dangerous conditions in your country with President Carter's emphasis on human rights? Three, the countries are worried that if they report poisonings, the FDA will start to check their food exports to the United States and find illegal residues. This could have a severe adverse impact on their export earnings."[17]

Poisons in a Coke bottle

PESTICIDE POISONINGS are much more common in the third world than in the industrial countries not only because of the more brutal working conditions there, but also because of hazards of distributing *any* poison in societies where most people cannot read and have never had to learn the dangers of manmade chemicals.

"Small shops in Indonesia sell pesticides right alongside the potatoes and rice and other foods," says Lucas Brader of the U.N. Food and Agriculture Organization (FAO). "The people just collect it in sugar sacks, milk cartons, Coke bottles—whatever is at hand."[18]

"The laws in less developed countries typically say no repackaging of pesticides," Fred Whittemore of AID explains. "But in the villages it is done routinely. Parathion in Coke bottles stuffed with newspapers with no label is typical."[19] Gramoxone, which contains the deadly weed-killer paraquat, is not only sometimes sold in Coke bottles—it's the same color as Coke.

In Pakistan and Middle-Eastern countries, peasants sometimes wrap pesticides in their turbans, then place the turbans back on their heads to carry the pesticides to the fields.[20]

"In the rainy season in many tropical countries, the plastic liners used in pesticide bags are used as raincoats," says Whittemore. "That is an acute problem causing poisonings."[21]

Gramoxone killed at least 18 people during a four-year

period in the Western Highlands of Papua New Guinea, where it is used on coffee plantations and home gardens. "On June 16 a pastor conducted a religious service at Tega village near Mt. Hagen. He accidentally gave gramoxone instead of wine for communion to four people. They all died over the next week," Dr. D. J. Wohlfahrt, assistant secretary for Health in the Mt. Hagen district, wrote in the Papua New Guinea *Post Courier* of July 25, 1980. "In mid-1979 a young father bought gramoxone and stored it in a bottle. He asked his young son to go and get him a drink. He accidentally brought back the gramoxone and gave it to his father. After a gulp, the father realized it was not water he had drunk. But it was too late—he died," Wohlfahrt says.

"Gramoxone is legally marketed by the manufacturer in plastic bottles with built-in carrying handles that are just perfect for villagers to store their drinking water in after they have used up the weed-killer.

"How many people are we prepared to kill for the convenience of also easily killing weeds?" asks the doctor.

Inadequate labeling or deliberate mislabeling of pesticides also causes poisoning in third world countries. During 1979 the government of Colombia fined Hoechst and Shell for mislabeling pesticides, and fined Dow, Velsicol, Ciba-Geigy, American Cyanamid, and Hoechst for selling substandard products.[22] A recent check in Mexico disclosed that more than 50 percent of the pesticides sold there were labeled incorrectly.[23]

"One aid post orderly came to collect his medicines at Mt. Hagen Hospital and brought an empty gramoxone bottle to put the cough mixture in. The label read 'Poison' and had all the instructions written in English, but how many plantation laborers or village people can read English?" asks Dr. Wohlfahrt.

"Disposal of pesticides is a major problem, too," says Virgil Freed. "One horrible example is dieldrin in the Cameroon. A couple of years ago too much dieldrin was ordered, and the extra drums were simply placed outside in a jungle area. Now the containers have deteriorated and the dieldrin is spilling all over. I was there and saw the chemical sitting in puddles on the ground. There were people living in huts nearby. There could very well be subtle effects on them."[24]

Indiscriminate, widespread promotion of pesticides is especially disastrous in the third world. In countries where most people cannot read, what use are warning labels on pesticide packages? In countries which outlaw unions that could protect farmworkers, what chance do peasants have against the crop duster's rain of poison? In countries with neither enough scientists to investigate pesticide dangers, nor enough trained government officials to enforce regulations, should foreign pesticide makers be given a free hand to push products so dangerous they are banned at home?

CHAPTER THREE

DUMPING: BUSINESS AS USUAL

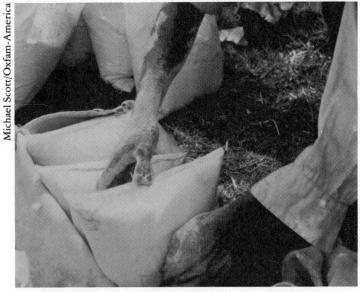

Michael Scott/Oxfam-America

Mixing Malathion powder for use in Sri Lanka anti-malaria project.

THE INTERNATIONAL TRADE in pesticides is an ever-changing affair. In the ebb and flow of commerce, buyers, sellers and products change constantly. To illustrate how hazardous pesticides find their way overseas, we have selected the example of DBCP, which provides a fascinating and well-documented case study of the problem.

Until recently, once or twice every working day a sealed semi-trailer drove through a grimy industrial section of the Los Angeles basin called City of Commerce. The truck moved slowly up Pacific Street past a row of dingy warehouses to a loading dock at the rear of Amvac Chemical Corporation's pesticide plant. There, from a storage area labeled "Restricted Area/Authorized personnel only beyond this point," light blue, 30-gallon drums stacked three high were loaded into the semi-trailer.

When it was filled, the rig headed for one of the interstate highways criss-crossing the area, and moved into the stream of traffic flowing back and forth across the country. The driver carried emergency telephone numbers and special instructions in case the colorless, odorless fluid in the drums somehow spilled or was released. No unloading or transfer of the toxic cargo was permitted en route to the shipping docks, where the chemical would be sent overseas.

The pesticide in the light blue drums was 1,2-Dibromo-

3-chloropropane, or DBCP. DBCP is a nematocide, effective against the small worms called nematodes which attack food plants such as pineapples, bananas and citrus fruits.

Many of the trucks carrying Amvac's DBCP were bound for the Gulfport, Mississippi, loading docks of Standard Fruit & Steamship Company, a subsidiary of Castle & Cooke. At Gulfport, the light blue drums were loaded onto a "banana ship," destined for Puerto Limón in Costa Rica, La Ceiba in Honduras, or Guayaquil in Ecuador.

DBCP bananas for U.S. tables

AFTER FOUR TO SEVEN days at sea, the banana ship docked and the pesticide drums were unloaded and taken to their ultimate stop—the vast banana and pineapple plantations of Castle & Cooke. Castle & Cooke is one of the largest foreign corporate landholders in Central America. Workers on the company plantations, mostly illiterate peasants, used this pesticide to kill soil-dwelling worms that attack bananas. Then virtually all the bananas were shipped to U.S. tables.

Castle & Cooke stopped importing DBCP from Amvac in late 1979 when the U.S. Environmental Protection Agency issued an emergency suspension of all uses of DBCP (except on Hawaiian pineapples) because DBCP is believed to cause cancer and make humans sterile.

"Castle & Cooke decided to stop importing DBCP from us directly," an Amvac marketing official explained. "Their policy is 'let's not cause any furor—we will get the stuff through local importers down there.' Now we have to contact the people who import it for them in Central America. Castle & Cooke won't buy it directly anymore, but they encourage their plantation managers to buy it from local importers down there."[2]

Castle & Cooke, however, maintains that it stopped purchasing and using DBCP, directly or indirectly, immediately following the EPA ban. According to Leonard Marks, Jr., executive vice president of the company, "It is Castle & Cooke's corporate policy that we will not use nor purchase product treated with any pesticide which is not specifically registered for that use by the U.S. Environmental Protection Agency."

Amvac, however, according to the marketing official, continued to sell DBCP after the ban "anywhere in the world where bananas, pineapples, citrus, and cotton are grown.[3] There's no problem with the ban of DBCP (within the United States)," he said. "In fact, it was the best thing that could have happened for us. You can't sell it here anymore but you can still sell it anywhere else. Our big market has always been exports anyway."[4]

Where giants fear to tread

RECENTLY, Amvac apparently suspended production of DBCP. For several years, however, it was the only company anywhere producing the substance.

When workers at Occidental's DBCP plant in California discovered in 1977 that many of them were sterile, the State of California banned the use of DBCP outright—an action that the federal Environmental Protection Agency waited two more years to take. Occidental, Dow and Shell quickly ceased their production. But little Amvac rushed in to seize the profit opportunities suddenly abandoned by the giants.

The company candidly explained its motives in its "10-K," an annual report required by the U.S. Securities and Exchange Commission: "Management believes that because of the extensive publicity and notoriety that has arisen over the sterility of workers and the suspected mutagenic and carcinogenic nature of DBCP, the principal manufacturers and distributors of the product (Dow, Occidental, and Shell Chemical) have, temporarily at least, decided to remove themselves from the domestic marketplace and possibly from the world marketplace," the report states.

"Notwithstanding all the publicity and notoriety surrounding DBCP," Amvac's report continues, "it was [our] opinion that a vacuum existed in the marketplace that [we] could temporarily occupy ... [we] further believed that with the addition of DBCP, sales might be sufficient to reach a profitable level."[5]

A former Amvac executive told us why company officials decided to produce DBCP. "They're not really for
spreading cancer or 'no-population' growth [a reference to
the chemical's sterility link]. But DBCP is very important to
them," he explained. "Quite frankly, without DBCP, Amvac would go bankrupt."[6]

Before moving into the DBCP business, Amvac was in
bad shape. In late 1978, one of its two principal stockhold-
ers, MTM Enterprises (controlled by Mary Tyler Moore
and her former husband, and which produces the popular
television show *Lou Grant*), decided to divest itself of its
12 percent portion of Amvac's common shares. Amvac
faced imminent collapse until its major creditors formed a
lenient repayment schedule, and even then the company
missed its debt payments several times.[7] DBCP offered
Amvac a way out of the financial doldrums; the company
chose to take it.

Amvac's yearly sales of roughly $10 million may be
dwarfed by the multi-billion-dollar behemoths of the
pesticide industry, but Amvac was not alone in profiting
from pesticide dumping. In fact, it shares its profits with
Dow, which received a three percent royalty on all DBCP
sold by Amvac under a patenting agreement.[8] Thus, even
though Dow no longer made DBCP it still profited via the
interlocking financial arrangements that link the small
companies to the large, giving all a share in the global pesticide business.

Velsicol's pesticide shell game

THE STORY OF the pesticide Phosvel—and the "Phosvel Zombies" it created—is a chilling example of the
cruel shell game the multinational companies can play as
they move their poisons from one country to the next,
trying to maximize sales before their pesticides are banned
again.

Phosvel, the exclusive brand name for an organophosphate nerve toxin called leptophos, was marketed by Velsicol Chemical Corporation, a division of the mammoth
Northwest Industries, which distributes everything from

Cutty Sark Scotch whisky to Fruit of the Loom under-wear.[9]

The dangers of Phosvel became public in 1976, when the U.S. Occupational Safety and Health Administration (OSHA) revealed that workers at Velsicol's Bayport, Texas, plant had developed serious disorders of the central nervous system. Fellow workers dubbed them "Phosvel Zombies" because they lost their coordination, and their ability to work, talk and think clearly. The workers sued Velsicol, and the company closed the plant.[10]

But even after all the publicity generated by the Phosvel Zombie scandal (including charges by U.S. Senator Edward Kennedy that the company knowingly continued to make Phosvel even after its employees became ill), Velsicol kept selling the pesticide overseas.

Velsicol's marketing of Phosvel shows that even when third world countries join the United States in banning especially dangerous pesticides, the multinational giants who control the global pesticide supermarket can often continue to sell dangerous chemicals for years.

Although more than four years have passed since Phosvel's dangers burst into the headlines, there is disturbing evidence that it is still on the market.

The EPA never allowed Velsicol to sell Phosvel in the United States, although it did routinely issue a one-year experimental use permit.[11] Velsicol used this permit to its advantage in Colombia. When the Colombian Committee for Environmental Information began campaigning against the pesticide's disabling side-effects, Velsicol first threatened to sue, then produced its experimental use permit as "proof" that Phosvel had been registered for use in the United States for eight years.[12]

Meanwhile, with $4 million in AID funds, 13.9 million pounds of Phosvel and other banned pesticides were shipped to 50 countries as part of the U.S. foreign aid program from 1971 to 1976.[13] (This practice was eventually stopped by a lawsuit brought by environmental groups.) In Egypt, a widely publicized Phosvel epidemic in 1971 killed over 1,000 water buffalo and an unknown number of peasants. The victims suffered a slow and agonizing death, gradually paralyzed until they asphyxiated.[14]

Velsicol beats the Phosvel ban

ALTHOUGH VELSICOL says it no longer manufactures Phosvel anywhere in the world, documents show that during 1978—two years after the notorious Phosvel Zombie publicity—Velsicol imported Phosvel into Costa Rica via three shipments originating in Panama and Mexico.[15] In addition, according to the government of Indonesia, large quantities of Phosvel are still being sold there.[16]

In Colombia, authorities banned Phosvel in July 1977.[17] Velsicol simply moved its stockpiles of Phosvel to a free trade zone, technically out of Colombian jurisdiction, and then shipped it to nearby countries where it was not yet banned.[18]

Attempts by the company to peddle Phosvel elsewhere, however, met with resistance.

After the Philippines banned Phosvel, according to that country's Pesticide Technical Services Chief Ricardo Deang, "Velsicol came to us and said, 'We want to export our stocks of Phosvel to Thailand.' But we couldn't let them do that to our sister country. So I said, 'You have to prove to us that Thailand really wants this stuff. Otherwise you must send it back to the States for disposal.' So then they ended up shipping it all back to the U.S."[19]

Guatemalan official Fernando Mazariegos remembers, "After the Phosvel scandal in the U.S., Velsicol came to us and said they wanted to study the possibilities for its use in Central America. We turned that request down. I think their purpose was to start selling a lot of it in Central America and in other countries."[20]

Company spokesman Richard Blewitt says Velsicol did not try to mislead officials in Colombia, the Philippines, or Guatemala about Phosvel. But he does admit that "what happened happened. We're trying [to] make sure that [it] never happens again. A new team has been amassed [at Velsicol]."[21]

Phosvel is not Velsicol's only hazardous pesticide. Once Phosvel was removed from the market, at least officially,

Velsicol began manufacturing EPN as a substitute. But EPN—now under EPA review—is believed to be twice as neurotoxic as Phosvel.[22] Velsicol also manufactures ingredients for three essentially banned organochlorines—heptachlor, chlordane and endrin—at a plant in Chicago, and formulates the finished products in Memphis. Most of the production is for the overseas market, since endrin use is severely restricted in the U.S., and heptachlor and chlordane are completely banned for agricultural purposes inside the continental U.S. (Like DBCP, endrin may be used on pineapples in Hawaii until the end of 1982—which testifies less to the need for either pesticide on pineapples than to the political clout of the Hawaiian Pineapple Growers' Association, dominated by Del Monte and Castle & Cooke.)[23]

Hooker: the Love Canal dumper

THE HOOKER Chemical and Plastic Corporation—infamous for the Love Canal tragedy—is another pesticide dumper. Hooker is a wholly-owned subsidiary of Occidental—one of the three major firms which ceased DBCP production when workers at its California plant discovered they were sterile.

At the Love Canal near Niagara Falls, New York, thousands of pounds of lethal chemical wastes at an abandoned Hooker chemical dumpsite percolated to the surface 20 years later. This tragedy is still being felt in the residential neighborhood today, four years since the site was rediscovered in 1976. An unusually large number of children are born with birth defects, adults and children are suffering from high rates of chemically-induced diseases, and a whole way of life has been disrupted as people have been forced to sell their homes to escape the leeching poisons. The Love Canal tragedy was a turning point in the growing movement of people fighting against the invasion of toxic chemicals.

In the third world, Hooker's marketing of pesticides may eventually cause similar tragedies. But third world peasants usually do not have access to information about

Hooker's toxic products and practices, and probably will not know what has affected them. In addition, most of them cannot simply move away like many Love Canal residents have done.

One example of Hooker's third world marketing practices occurred in 1976. Hooker voluntarily withdrew its EPA registration of the organochlorine BHC, after feeding tests with mice showed it causes tumors, kills fetuses and causes premature births, and has other dangerous reproductive effects even when absorbed in tiny concentrations. But when it withdrew BHC from the U.S. market, *Hooker explicitly stated that it would continue producing the chemical for use overseas.*[25]

In recent years, records indicate, Ortho has imported BHC into Costa Rica,[26] the German firm Schering imported BHC into Colombia,[27] and, according to U.S. Department of Agriculture cables, BHC was used on coffee grown for U.S. consumption in Peru and Guatemala.[28]

CHAPTER FOUR

THE PESTICIDE BOOMERANG

Pesticide barrels in Guatemala.

PESTICIDE POLLUTION does not respect national borders. As one of the world's largest food importers, we in the United States are not escaping hazardous chemicals simply by banning them at home. (See Table Two.)

Approximately 10 percent of our imported food contains illegal levels of pesticides, according to the U.S. Food and Drug Administration (FDA).[1] But that 10 percent is deceptive. The FDA's most commonly used analytical method does not even check for 70 percent of the almost 900 food tolerances for cancer-causing pesticides.[2] (A tolerance is the amount of a pesticide allowed in any particular food product.)

In addition, the FDA frequently finds mysterious, unknown chemicals in imported foods. Government investigators believe that some of these fugitive chemicals come from the millions of pounds of "unregistered" pesticides the EPA allows U.S. manufacturers to export without divulging any information about their chemical makeup or their effects on people or the environment.

Knowing how little we know, we suspect these statistics

from the General Accounting Office (GAO) represent only the tip of the iceberg:

• Over 15 percent of the beans and 13 percent of the peppers imported from Mexico, during one recent period, were found to violate FDA pesticide residue standards.[3]

• Nearly half the imported green coffee beans contain levels (from traces to illegal residues) of pesticides that are banned in the United States.[4] (See Table Three.)

• Freshly cut flowers flown in from Colombia caused a rash of organophosphate poisonings among American florists.[5]

• Imported beef from Central America often contains pesticide contamination. The GAO has estimated that 14 percent of all U.S. meat is now contaminated with illegal residues,[6] and imports make a significant contribution to that total.

The pesticide residue problem has escalated to such a level that all beef imports from Mexico,[7] El Salvador,[8] and Guatemala[9] have been halted by the USDA. Agricultural practices in those countries, including heavy pesticide use on crops next to cattle-grazing land, have backfired on ranchers raising beef for the U.S. market.

Despite the widespread contamination of imported food, FDA inspectors rarely seize shipments or refuse them entry. Instead, a small sample is removed for analysis while the rest of the shipment proceeds to the marketplace . . . and the consumer. The rationale is that perishable food would spoil if held until the test results were known. But by the time the test results *are* available—showing dieldrin or parathion or DDT residues—the food has already found its way into our stomachs. Recalls are difficult.

During one recent 15-month period, government investigators found that *half* of all the imported food identified by the FDA as pesticide-contaminated was marketed without any penalty to the importers or warnings to consumers! Even products from importers with repeated violations were routinely allowed to pass. Some examples:

• USDA officials in Dallas noticed a strong "insecticide-like smell" in a batch of imported cabbage from an importer with a record of shipping contaminated products.

Despite USDA's complaint, the FDA allowed the cabbage to go to market. A sample that had been removed for testing later revealed illegal levels of BHC, the dangerously carcinogenic pesticide whose registration was cancelled in 1976 at Hooker Chemical's request. But it was too late to recall the cabbage.[10]

• Peppers from a shipment that was sent on to supermarkets turned out to have 29 *times* more pesticide residue than allowed by U.S. law.[11]

In a world of growing food interdependence, we cannot export our hazards and then forget them. There is no refuge. The mushrooming use of pesticides in the third world is a daily threat to millions there—and a growing threat to all consumers here. Therefore we and third world people are allies in a common effort to halt the production of hazardous pesticides and contain *all* pesticide use to safe levels.

CHAPTER FIVE

PESTICIDES TO FEED THE HUNGRY?

Martin Walterding

Departmento Launion, El Salvador. After the cotton harvest, cattle are herded into the fields to graze on the contaminated vegetation. Beef from the herd is both consumed locally and exported to the U.S.

WE SEE NOTHING wrong with helping the hungry world eat," says an executive of the Velsicol Chemical Company, defending his company's overseas sales of Phosvel after it was banned in the United States.[1] And many would agree with his logic: since we need pesticides to produce more food for the hungry, pesticide dangers are a necessary evil—part of the price of averting famine. "Men will not starve because there are hazards in killing pests," is the way a Rohm & Haas official makes the same point.[2]

But in the course of our investigation, we came to a startling conclusion: over half, and in some countries up to 70 percent, of the pesticides used in underdeveloped countries are applied to crops destined for export to consumers in Europe, Japan and the United States.[3] The poor and hungry may labor in the fields, exposed daily to pesticide poisoning, but they do not get to eat many of the crops protected by pesticides.

In Central America a staggering 70 percent of the total value of agricultural production—mainly coffee, cocoa and cotton—is exported, despite widespread hunger and malnutrition there.[4] Cotton is one of the biggest pesticide users. In tiny El Salvador, cotton production absorbs one-fifth of all the deadly parathion used in the *world*.[5] Twenty-four hundred pounds of insecticides are used each year on every square mile of cotton fields in the country.[6] Yet cotton contributes to the global food supply only in processed cattle feed for Latin America's burgeoning beef production, almost half of which is exported to the United States and Europe.[7] The meat remaining for local consumption is eaten by the rich and the middle classes, not by the hungry.

Herbicides like 2,4,5-T and 2,4-D (the basic ingredients of the infamous Agent Orange) are also used to help clear huge amounts of forest for grazing land in Latin America. The herbicide 2,4,5-T leaves residues of its contaminant, dioxin, in soil and water. Dioxin, one of the deadliest poisons ever developed, shows up later in birth defects, skin rashes and miscarriages.

In Indonesia, estate-style farms growing export crops— coconuts, coffee, sugar cane and rubber—consume 20 times the quantity of pesticides used by the small holders growing food for local markets. This, despite the fact that small holders cultivate seven times more acreage than the estates.[8]

Some might argue that although export crops do not directly feed hungry people, at least the foreign exchange earned benefits them indirectly: it is used to import economic necessities for development. But even the most superficial look at development in most third world countries belies this assumption. Foreign exchange earned by agricultural exports does not return to improve the lives of the workers through better wages, housing, medical care, or schools. Instead the foreign exchange is most often plowed into luxury consumer goods, urban industrialization, tourist facilities, and showy office buildings—all geared to the budgets and tastes of the top 10 to 20 percent living in the cities.

The perfect banana

ONE REASON pesticide use is so much more intense on export crops than on subsistence food crops is that the multinational corporations which control the production and marketing of exports demand a blemish-free product. Nothing less, they say, will meet the discriminating standards of the consumers in Europe, North America or Japan.

"The Japs eat with their eyes" is how the manager of a Philippine banana plantation explained why they went to such lengths to produce a blemish-free fruit to ship to Japan.[9] In the United States, too, it is estimated that 10 to 20 percent of insecticides used on fruits and vegetables serve only to improve their appearance.[10]

Most people think of multinational food corporations in the third world as big plantation owners. But over the last 20 years, corporations have become leery of owning land directly. As the U.S. Overseas Private Investment Corporation warns, the possibility of "expropriation, revolution or insurrection [makes] plantations a poor risk."[11] Multinational food producers and marketers such as Del Monte, United Brands (formerly United Fruit), and Castle & Cooke (Dole brand) have hit upon a safer strategy—contract farming. Rather than own land directly, these companies now often contract with large local landowners to produce crops for export to consumers in the industrial countries.

A contract farming boom hit southern Mindanao, the Philippines, in the late sixties. Before that time there were no bananas growing on its rich coastal plains. Small farmers and tenant farmers grew rice and abaca. Then came the multinational corporations, seeking contracts with local entrepreneurs to produce bananas for the lucrative Japanese market. Within ten years the entire area was transformed: now 21 giant plantations cover 57,000 acres, and bananas have become one of the country's top agricultural exports.[12] In order to fulfill their banana contracts, the local entrepreneurs had to push small holders, tenants and "squatters" off the land. (Some of the so-called squatters had worked the land for more than a generation.)

Although the multinational corporation may not own the land, it still calls the shots. When the corporation signs a local entrepreneur under contract, it specifies not only the amount of fruit or other commodity to be produced but also the amount of fertilizers and pesticides to assure high yields and blemish-free products.[13]

Lifetime debt to pesticide companies

ONCE LOCKED INTO the banana export contract, the plantation owner is totally dependent on the multinational firm. "Money is deducted from the banana grower's earnings to pay for things like pesticides and irrigation," explains Father Jerome McKenna, a U.S. missionary who worked in the area. "It's part of the contract. Those banana growers will be in debt to the pesticide companies for the rest of their lives."[14]

Typically pesticides are applied at three stages in the banana production process. Workers with heavy tanks strapped to their backs (and no masks or protective covering) routinely spray every tree. Twice a month a pesticide plane passes over the plantation, blanketing everything, including the drinking water supply. A group of banana workers recently petitioned Castle & Cooke to stop heavy pesticide spraying after local studies showed that the workers have dangerously low oxygen levels in their blood, making them more susceptible to disease.[15]

In the packing sheds, the bananas are dumped in long water-filled troughs to remove some of the pesticides. "What bothers me most," says McKenna, "is that these people have very little protection from the chemicals they come in contact with. The women have their hands in the water up to their elbows all day long. They don't wear any gloves. Their only protection is plastic-type aprons they fashion for themselves."[16] Finally, to protect the fruit during its long ocean voyage, women workers in the packing sheds spray every bunch of bananas with a fungus-killing agent.[17]

McKenna checked at two nearby hospitals for reports of pesticide poisonings. One, run by Castle & Cooke, "didn't have any cases." But the other hospital, run independently of the company, had "reports all around of people poisoned by pesticides."[18]

The contract farming system also gives the multinationals an easy way to avoid responsibility for pesticide poisoning. They can simply blame the local plantation owner for being careless.

The examples of cotton in El Salvador or bananas in the Philippines tell us that, in large measure, pesticides in the third world actually feed the well-fed, but endanger the poor and the hungry. Since the mid-50s, the growth rate of export crops—which receive the overwhelming bulk of pesticides—has exceeded that of food crops.[19] Between 1952 and 1967, for example, cotton acreage in Nicaragua increased fourfold while the acreage in basic grains was cut in half.[20] Thus it is hardly surprising that the demand for pesticides in the third world has soared. What is surprising is how many believe that their principal use is to save crops to feed the hungry.

More food and yet more hunger

WHILE IT IS TRUE that most pesticides in the third world are used on luxury export crops, in the last 20 years third world farmers growing basic food crops—especially rice and wheat—have also been encouraged to use ever greater quantities of pesticides. As part of the "green revolution," hybrid seeds were developed which produce higher yields, given the correct amount of fertilizer and water; but the hybrids are much more susceptible to pests. Bred in the laboratory and in test fields in a foreign setting over only a few years, these "miracle seeds" do not have the pest resistance characteristic of traditional seeds, bred over thousands of years in the same locality in which they are used.[21] To make up for this vulnerability, the new seeds must be protected with more pesticides.

Throughout much of the third world, international lending agencies and government development programs have encouraged the use of these new seeds, often making their use a condition for receiving farm credit.[22] Once third world farmers begin using the new, more vulnerable seeds, they have no choice but to vastly increase their use of pesticides.

Few dispute that the new seeds and their accompanying inputs—fertilizers and pesticides—have increased grain production, notably in Asia. But growing more food doesn't necessarily mean alleviating hunger. What we have learned is that food production can increase while the poor majority gets even more hungry.

Take the Philippines. It is the home of the prestigious International Rice Research Institute which helped instigate the "green revolution" in Asia. During the 1970s, use of the new seeds spread throughout the country. Accompanying their proliferation, pesticide imports leapt fourfold between 1972 and 1978.[23] As a result of the new seeds and new inputs, rice production almost doubled in the Philippines in little more than a decade.[24] Indeed, in the late 1970s, the Philippines became a rice exporter. But has this production success reduced the hunger of the Philippine poor? No. According to studies by the Asian Development Bank and the World Health Organization, Filipinos are now the worst fed people in all of Asia, with the exception only of war-torn Kampuchea.[25]

How can there be more food produced and yet greater hunger? The answer is that the green revolution strategy for producing more food forces more and more people off the land. Mechanization robs them of work. Dependency on irrigation, pesticides and fertilizers—all required by the new seeds—favors the wealthier, literate farmers who have access to credit and political pull. Without land to produce food or money to buy it, people go hungry no matter how much their country produces.

This dramatic transformation is documented in the International Labor Organization's study of rural poverty. After studying seven Asian countries, comprising 70 percent of the rural population in nonsocialist underdeveloped countries, the ILO reported that the rural poor have become measurably poorer than they were 10 or 20 years ago. The study concludes: "The increase in poverty has been associated not with a fall but with a rise in cereal production per head, the main component of the diet of the poor."[26]

Another ILO study of the "green revolution" points to

vast increases in wheat yields in the Punjab district of India in the 1960s. Yet simultaneously, the portion of the rural population living below the poverty line increased from 18 to 23 percent.[27] "Economic prosperity has not simply missed these people," the study concludes. "Their ability to supply their own basic needs has been gradually but unrelentingly reduced...."[28]

The poor: not a lucrative market

T HE NARROW PRODUCTION push embodied in the "green revolution" strategy, helping to enrich the well-placed farmers and further impoverish the rural poor, has itself encouraged the shift toward export crop production that we discussed above. This is true in part because impoverished people simply do not make up a lucrative market. So, as in the Philippines, a staple food like rice is exported while Filipinos—without money enough to buy the rice—go hungry. Or production shifts from staple foods needed by the poor and toward luxury items demanded by the rich. Corn and bean production in Mexico, for example, has declined while production of luxury fruits and vegetables for the U.S. market and feedgrains such as sorghum have greatly increased. Almost 32 percent of basic grain staples are now fed to livestock in Mexico.[29] In Brazil the figure is 44 percent.[30]

Thus the rationale of using more pesticides to protect crops to feed the hungry simply does not hold up. First, we discover that most pesticides are not used to protect food crops anyway! Second, pesticides to protect the more vulnerable grain seeds of the "green revolution" are part of a production strategy benefiting the better off. While increasing production, this strategy cannot eliminate hunger because it fails to address the question of *who controls* that production. Under these conditions, the extra food which pesticides help to grow is frequently either eaten by the better off, exported or fed to livestock. The whole equation bypasses the fundamental problem: the hungry have neither money to buy food nor land to grow it on.

CHAPTER SIX

THE GLOBAL PESTICIDE SUPER-MARKET

Guatemala News & Information Bureau

Pesticides on shelf of store in Guatemala.

FROM THE BILLBOARDS of rural Nebraska to shanty-town walls in Kenya, pesticide company advertising is part of the scenery. The language may be English or Spanish or Swahili, but the message is the same: you need our brand of pesticide if you want a good crop.

"Whenever a new pesticide hits the area, every farmer knows about it right away," says Dr. Lou Falcon, a University of California entomologist who has studied Central America. "There is heavy publicity by the companies—big billboards, radio and newspaper ads." [1]

Using sophisticated marketing techniques and their worldwide network of subsidiaries and affiliates, the giant multinational pesticide manufacturers—such household names as Dow, Shell, Chevron, Bayer, Dupont—have created a global supermarket, its shelves stocked with products so dangerous they have been banned in the countries where they have been investigated.

As we have said, the multinationals claim they sell pesticides overseas merely to supply a demand, a demand for their products to help feed a hungry world. But the fact is that multinational companies use sophisticated mass marketing techniques to *create* a demand in the third world.

"Those pesticide boys are all over the place down there," says Michael Moran of the Interamerican Institute for Agricultural Sciences in Costa Rica. "It's amazing how they get down to the grass roots. Very few places are left in Latin America which are in isolation from the new technologies, including pesticides."[2]

"We have overseas offices in almost every country in Asia," explains René Montmeyor, an agricultural product supervisor for Stauffer Chemical Company. "We have exclusive distributorships in most of those countries, too. We have our technical people who instruct farmers how to use our pesticides."[3]

Ads for pesticides appear prominently in third world agricultural journals. Away from the eyes of U.S. regulators, pesticide companies often extol the virtues of pesticides banned in the U.S.

At a supply center for the Kenyan Farmer's Association in Nairobi, a reporter spotted aldrin, BHC and chlordane—all banned from most uses in the United States—for sale on shelves and listed in the association's inventory. They were being sold by local subsidiaries of European pesticide companies—ICI, Bayer and Shell.[4]

Formulating their way around regulation

TO ESCAPE REGULATION in their home countries, the multinationals have discovered a clever strategy: they simply ship the separate chemical ingredients of a banned pesticide to a third world country, then manufacture it there in "formulation plants." From the third world country, the prepared pesticide can often be re-exported to any third country, free of regulation.

"It's a real Mafia-type operation," says Dr. Harold Hubbard of the U.N.'s Pan American Health Organization. "Global companies are setting up formulation plants all over the world. [They] simply go into less developed countries, give a banned pesticide a local name, and then turn around and sell it all over the world under that new name."[5]

"Formulators buy basic ingredients from importers and then put them together and call the product a name like 'Macho' and say it will kill anything," explains Frank Penna, a consultant to the Policy Sciences Center. "Usually it ends up killing the farmer."[6]

("Macho" competes with other chemical weapons with such names as Ambush and Fumazone to battle an army of enemies led by kernel smut, the stinkbug, the whorl maggot, and the black whip and tip smut.)

The pesticides are dangerous before they ever reach the fields. A plant in Kenya which formulates BHC provides no protection for the workers mixing the chemicals. "The workers' eyes were all sunken, and they looked like they had TB," says a University of Nairobi professor who visited the plant. "There are regulations against this sort of thing, but there is no manpower for enforcing the regulations. And no one complains. The workers are perfectly happy until one of them gets sick, and then he's just fired."[7]

In Latin America, "you can see the dust rising from those formulation facilities for miles," says AID's Whittemore. "I wouldn't dare walk into some of them. There are no decent health or environmental standards for most of them—it's a terrible problem."[8]

The worst formulators, Penna says, are the "pirate operators—little whiskey-still-like operations." An estimated 8,000 of them have opened in Brazil alone.[9] But the large-scale formulation plants are foreign-owned.

Like many other third world countries, Brazil offers special incentives to bring foreign chemical plants into the country: deferral of taxes, exemption from import duties, government-sponsored clearing of land for the plants.[10] Shell has put $20 million to $30 million into new plants under these incentives over the past few years. Dow has a 2,4-D plant there.[11] The Swiss firms Sandoz and Ciba-Geigy set up a joint operation.[12] And the largest pesticide company in the world—Bayer—has formulation plants in Brazil as well as in virtually every other country with a market large enough to warrant one. (See Table Four.)

Formulation plants are also spreading throughout Asia:

India. Many pesticides that have been banned or heavily restricted in the United States are produced in India, including BHC and DDT.[13] Union Carbide, ICI, Bayer, and Hoechst have plants there.[14]

Malaysia. Dow and Shell alone formulate one-quarter of all liquid pesticides here. Three organochlorines banned in the United States—aldrin, DDT and BHC—constituted 730 of the 960 tons of pesticides manufactured in Malaysia in 1976.[15]

Indonesia. Bayer, ICI, Dow, and Chevron dominate the local pesticide manufacturing industry, accounting for over 70 percent of the total production in 1978.[16]

This trend toward formulation plants is paralleled in many heavily regulated industries which are also moving their production facilities overseas.

Seeds: the final round?

THE MULTINATIONAL pesticide producers already control the manufacturing, distribution and promotion of pesticides at the global supermarket. Now they are working on a strategy to control an even more basic agricultural "input," the seeds themselves.

"Where might a chemical company interested in agricultural chemicals go?" rhetorically asks a high official of the Chemical Manufacturers Association. "Obviously, into seeds," he answers. "Some members of the chemical industry are getting into seed development."[17]

The FAO estimates that by the year 2000, 67 percent of the seeds used in underdeveloped countries will be the "improved" varieties, which in most cases are more vulnerable to pests.[18] Since virtually all pesticides are produced in the industrial countries, that means more pesticide exports to the third world.

For the agri-chemical multinationals, plant patenting provides greater inducement to add seeds to their conglomerate families. Championed by the American Seed Trade Association and the USDA, controversial legislation to allow the patenting of all U.S. crop varieties has been debated in Congress since early 1980.[19] The bill would extend the patent umbrella to six crop varieties that were excluded from the original 1970 plant protection act.[20]

Already a few multinational corporations, many of them pesticide producers, control the seed patents for several important crops. Of the 73 patents granted for beans, for example, over three-quarters are held by just four corporations: Union Carbide, Sandoz, Purex, and Upjohn.[21] Two Swiss-based companies, Sandoz and Ciba-Geigy, alone control most of the U.S. alfalfa and sorghum seed supply.[22]

Chemical companies are buying traditional seed supply firms, and their patentable "commodities," at an alarming rate. After the first wave of acquisitions, the international pesticide giants Monsanto, Ciba-Geigy, Union Carbide, and FMC are ranked among the largest seed companies in the United States.[23] (See Table Five.) Between 1968 and 1978, multinationals—mainly chemical and pharmaceutical companies—bought 30 major seed companies. Today, the largest seed enterprise in the world is Shell, the oil and petrochemical giant which controls 30 seed outfits in Europe and North America.[24]

Entering the $10-billion-a-year seed industry is a natural for the multinational pesticide producers. They already have the marketing and distribution structures for reaching the small farmer throughout the world, explains *The Global Seed Study,* a $25,000-a-copy investment guide sold to potential seed investors.[25] The study points out how seeds and chemicals can work together, as in the possibility of "seed coatings and pelleting, utilizing the seed as a delivery system for chemicals and biologicals to the field."[26]

By cornering the global seed market, the companies apparently plan to insure that farmers the world over are dependent on their seeds, as well as their fertilizers and pesticides.

"Obviously they're being damn quiet about it," says an industry official. "But some of those high yield seeds require particular applications of fertilizers and pesticides to produce their high yields." [27]

Now that the chemical companies have entered the seed business, they hold the enviable economic position of helping to aggravate the (pest) problem for which they also offer their (chemical) cure. If the chemical industry's monopolization of the world's seed stock is successful, we will be one critical step closer to the ultimate corporate vision of the global supermarket, where every grower in the world is hooked on patented seeds and the pesticides they require.

Genetic uniformity

COMMON NON-PATENTED varieties often become extinct and disappear as seed varieties are patented. By 1991, the FAO's Erna Bennett estimates, three-quarters of all vegetable varieties now grown in Europe will be extinct due to patenting, which is more advanced in Europe than in the United States. [28]

As fewer seed varieties are used to grow larger crops, the earth's genetic base is narrowing. [29] At the same time, the uniform high-response variety seeds of the green revolution are displacing centuries-old varieties and accelerating their disappearance from the earth's seed stocks.

The implications of this genetic uniformity may be devastating for our food supply. The hybrid, high-yielding seeds do not have an inbred resistance to pests and are usually planted in huge fields that can satisfy swarms of the same type of pest. "If the crop is a monoculture, you no longer have the buffers of different varieties of crops," adds a congressional aide working on the plant patenting issue. "What you've got instead is a super-highway for these insects." [30]

Scientists now suggest that genetic uniformity was the underlying cause of the Irish potato famine in the late 1840s. Then, a single potato variety imported from the Caribbean was struck by blight and over one million people starved to death. [31] More recently, the United States had a glimpse of what this genetic uniformity means, when 15 percent of the nation's corn crop was destroyed by a pest epidemic in 1970.[32] (Only six seed types make up 71 percent of the domestic corn crop.[33])

The world's farmers will become even more dependent on pesticides as they find that their seed varieties are less able to resist the diseases and pest epidemics that sweep through local areas periodically.

CHAPTER SEVEN

LUBRICATING THE SALES MACHINE

This store in Usulatan, El Salvador, sells pesticides for agricultural use.

Martin Walterding

O N A HOT SUMMER day in the rice paddies of Pakistan, a portly man's loose-fitting garments flap softly in the breeze as he makes his daily rounds by motorcycle over roads too rough for automobiles. The man is an agricultural credit officer from the National Bank of Pakistan. His job is to visit remote villages, offering loans to farmers to buy the technology of modern agriculture. With the loans (averaging $187),[1] farmers will be able to buy government-subsidized pesticides, including dieldrin, endrin, heptachlor and BHC—all banned in the U.S.[2]

In the plush conference room at the World Bank's headquarters in Washington, D.C., top-level officials are discussing the merits of a massive development loan to Brazil which will include huge purchases of "ag-chemicals" to increase yields of a new export crop.[3] The same pesticides banned in the U.S. and available in Pakistan are also sold in Brazil.[4]

Off the western coast of Africa, two newly independent islands, São Tomé and Principe, are attempting to convert former Portuguese plantations to diversified food crop production. To clear the land, they have applied to the U.S.

Agency for International Development (AID) for help.
One of the four herbicides AID recommends for the job is
paraquat, manufactured by Chevron.[5]

Paraquat, notorious as a dangerous contaminant of Mexican marijuana, is currently under review by the Environmental Protection Agency for possible new restrictions in
the United States because it is increasingly being used to
commit suicide. There is no known antidote. "The hazard
in paraquat is that it might be gotten by third parties," says
AID's Fred Whittemore. "Once they take it they have from
12 to 20 days to live and that's that. It must be more or less
kept under lock and key with very strict controls. As long
as that's the case we okay its use."[6]

In the large, well-irrigated fields of Argentina, the International Institute of Agricultural Sciences (IICA) is helping
to devise a system to "improve the efficiency of the distribution of agricultural products and inputs."[7] Until recently
an arm of (and still closely allied with) the Organization of
American States, the Institute maintains offices in every
Latin American country. It supplies technical assistance for
agricultural marketing development at the village level.
"We're trying to link marketing and distribution with the
grower,"[8] remarks Michael Moran, the Institute's North
American representative.

Working on completely different levels, these four organizations—the National Bank of Pakistan, the World
Bank, AID, and IICA—are all integral parts of a complex
network of "facilitators" for worldwide pesticide trade. In
the lingo of the agricultural economists who chronicle this
push toward agricultural "development," these institutions
help "synch" (synchronize) third world farmers into the
global marketplace.[9] This is accomplished through two
commodities controlled by powerful institutions in the
industrial countries—*money* and *information*.

The capital invasion

AS THEIR RELIANCE on third world markets grows,
the multinational chemical producers face a vexing
problem: most of the world's farmers cannot afford high-

technology "inputs" such as pesticides, fertilizers, tractors and high yield variety seeds. Conveniently, world bankers—who work closely with the multinationals—have stepped in, and mobile banks have begun to appear in the third world.

From Pakistan to Africa, vans and motorcycles pull into marketplace squares in small villages. Officials in the vans arrange credit for local farmers and then drive to the next village. They are local representatives of the national or regional development banks, the provincial fingertips of grand development schemes designed to integrate the farmers more completely into the global supermarket. Credit is the critical link.

The arms of the international credit system are the nine regional development banks, which supplement the World Bank's global operation. Reaching into Asia, Africa, and Latin America, their credit programs play an important role in providing capital to third world farmers to purchase pesticides and other agricultural implements. A study sponsored by the U.N. Environment Program (UNEP) in 1979 criticized the regional banks for completely ignoring the environmental impacts of loan projects.[10] Spurred by UNEP, the regional banks then agreed in principle to a protocol for environmental impact review of all projects. But implementation would require the hiring of a consultant to oversee the loans; at press time no hiring date had been set.[11]

If the mobile village banks are the fingertips, and the regional banks the arms, then the World Bank is indisputably the head of the international lending body. The Bank provides over one-quarter of the aid loaned by all multilateral sources, with 2,100 projects in 116 developing countries.[12] About 25 percent of all World Bank loans go to agricultural projects. Most involve irrigation, seed improvement, chemical fertilizers and pesticides.[13]

Incredibly, the Bank's staff does not include a single pest control expert to advise on pesticide use in its agricultural projects.[14] "Loans are too big and too insensitive to take the proper use of pesticides into account," one Bank official admits. "What happens a great deal is that the Bank will give a bulk loan to a government for 'chemicals' and not specify the specific pesticide."[15]

Uncle Sam's credit guarantees for multinationals

AID FINANCED massive overseas shipments of banned pesticides until a 1975 lawsuit brought by environmental groups ended the practice.[16] AID still finances pesticides exports to the third world.

Besides AID, two other U.S.-government-sponsored finance and insurance agencies have helped quadruple U.S. exports to the third world since the late 1960s. They are the Export-Import Bank and the Overseas Private Investment Corporation (OPIC). The Eximbank offers direct financing. Both agencies provide "political risk insurance" that guarantees U.S. corporations against losses due to war, revolution, insurrection, expropriation, or currency inconvertibility for as long as 20 years. The U.S. Treasury stands ready to cover the company's losses. Political risk insurance of this type is a key ingredient in the financial formula required by multinationals when making overseas investment decisions.

OPIC's coverage in recent years has gone mainly to a small number of huge corporations operating in a handful of third world countries with authoritarian regimes. Between 1974 and 1976, a full 60 percent of OPIC insurance went to six countries—Brazil, the Philippines, South Korea, Indonesia, Taiwan and the Dominican Republic.[17] Labor unions have also criticized the agency's activities as promoting the "runaway shop" movement into low-wage areas.[18]

OPIC has singled out the chemical industry for special largesse: *three of the four top recipients of OPIC support in recent years were large chemical producers.* Between 1974 and 1976, Dow topped the list, receiving $181 million in U.S. taxpayer guarantees; W. R. Grace received $70 million.[19]

The knowledge monopoly

A CALIFORNIA FARMER complained at a "Politics of Pesticides" conference in Berkeley early in 1980: "If

I only had access to information on alternatives to pesticides, I wouldn't be forced to use them."[20] The only pest control advice he gets is from pesticide salesmen or from local government agricultural officials. The government agents often cooperate closely with the industry. In fact, in California the vast majority of local pest control advisors received some form of financial sponsorship from the pesticide industry, according to a recent California Department of Food and Agriculture study.[21]

But however scarce, information on pesticide alternatives is far more available in California—the most intensive consumer of pesticides in the world[22]—than in the third world.

The late entomologist Robert van den Bosch dubbed the small group of pest control experts, company salesmen, and government officials the "Pesticide Mafia."[23] Notable in this circle are scientists from land-grant universities (the University of California is a prime example) who advise farmers and governments while doing their research with funds from pesticide corporations and testifying in court on behalf of those same firms.[24]

On the international level this Pesticide Mafia has infiltrated the United Nations and its many organizations.

The U.N.'s FAO as industry broker

THE UNITED NATIONS' Food and Agriculture Organization (FAO) was created to provide an independent pool of experts to consult with underdeveloped countries on a wide range of issues, including pest control. The major goal was to overcome world hunger. But under relentless pressure from industry, the FAO has often operated as a critical link between underdeveloped nations and multinational agribusiness firms.

Until 1978, this link was institutionalized in the Industry Cooperative Program (ICP), a "non-profit" organization of agribusiness corporations which was granted a secretariat and official status by the FAO in 1966. For 12 years, the industry most threatened by non-chemical alternatives to pesticides was able to work in close collaboration with the

FAO. A key component of the ICP was the Pesticide Working Group, whose members included BASF, Bayer, Ciba-Geigy, American Cyanamid, FMC, Hoechst, Hoffman-LaRoche, Imperial Chemical, Sandoz, Shell, and Stauffer chemical companies.[25]

The Industry Cooperative Program itself declared, under the FAO letterhead, that its "primary objective is to stimulate agro-industrial expansion in developing countries . . . [and] to demonstrate that far-sighted and responsible international business contributes to social and economic development by means of fostering profitable enterprise."[26]

ICP participation in the FAO inevitably led to direct industry involvement in FAO pest control programs. Hoechst, for example, became an advisor to the Tanzanian government on insecticides and spraying equipment, using a government agricultural extension officer to supervise the spraying. Hoechst was even given the power to fire an extension officer who did not supervise "properly."[27] And according to a former U.N. official working in Bangladesh, the U.N. field representative in that country was an executive of a large European chemical company which was also a main supplier of malathion for the "malaria eradication" project there.[28]

An example of the industry's influence on the FAO is contained in one of the Pesticide Working Group's action papers, printed on U.N. stationery. It was presented to the 1974 World Food Conference in Rome. Pesticides are necessary to solve the hunger problem in underdeveloped countries, stressed the document. It urged that public funds be used to establish an international stockpile of "essential pesticides" (DDT included).[29] Corporations should work more closely in training government technical staffs in pesticide use, according to the paper. And the approval of new pesticides should be expedited.[30]

Three years later, the FAO Plant Protection Service presented a comprehensive plan for "provisional" pesticide registration. The plan opens the door to hazardous pesticides by permitting the limited use of certain pesticides, even though not all health and environmental safety data is yet available.[31]

By the mid-1970s, critics began to attack the cozy FAO/industry relationship. "The cooperative relationship simply became politically untenable," recalls Don Kimmel, the FAO's North American liaison officer.[32] Political pressure increased to the point that ICP was kicked out of the FAO in 1978.

The cozy relationship continues.

LESS THAN A YEAR after ICP's ejection from the FAO, an ostensibly new group, the Industry Council for Development (ICD) was formed to pick up where the ICP left off. Although not formally affiliated with the FAO, the Council has actually expanded the functions of the old ICP by establishing strong, informal ties with all of the U.N. agencies concerned with development and agriculture.

Funded by 32 agribusiness multinationals, the Industry Council for Development has offices in the United Nations Plaza in New York. The two chief officers of the new group were also key figures in the old: Walter Simons, executive director, was director of ICP; and J. I. Hendrie of the Shell Chemical Group, in charge of the seed development program, served as director of the ICP's environment subcommittee.

"Conceptually ICD is still the same as the old ICP," Walter Simons admits. "Only now ICD is officially independent of the U.N. It also offers its services to many different multilateral agencies, including development banks, AID, and UNDP."[33]

The FAO's Kimmel remarks, "They [ICD] are interested in technology transfer.... They put out a lot of fancy materials—it's meant to make you buy chemicals."[34]

A supermarket of agricultural development

KIMMEL DESCRIBES the FAO itself as a "supermarket of agricultural development."[35] Through its connec-

tion with the Industry Council for Development and other
corporate groups, the FAO has often served as another link
in the chain pulling third world farmers into the corporate
network. The FAO publishes guides to pest control research projects around the world. It recommends to farmers which pesticides to use for particular pest problems.
One FAO study cited malathion and lindane—restricted in
the United States—as the most effective pesticides for
stored-grain pests.[36]

Other facilitators of sales in the global pesticide supermarket include the nine "green revolution" research institutes in the third world responsible for developing "miracle seeds" and adapting them to local crops and climatic
conditions. Their research projects are often closely coordinated with agribusiness' need to sell chemical pesticides, fertilizers and machinery.

One of the most prominent of the nine is the International Rice Research Institute (IRRI), based in the
Philippines. "We work pretty closely with the IRRI in developing new herbicides," says a Stauffer Chemical research official. "IRRI is doing a fantastic job. We're defi-
nitely interested in their work on herbicides and insecticides—like breeding rice resistant to an herbicide."[37]
Stauffer and many other chemical companies regularly do-
nate money to IRRI and the other international research
institutes.

International banks, U.S. government agencies insuring
corporate investments, a chemical industry promotion arm
working closely with the U.N., and international agricultural research institutes—all these institutions comprise a
network contributing to indiscriminate pesticide use in the
third world and dependency on imported, often hazardous
chemicals.

CHAPTER EIGHT

WITH THE ADVICE AND CONSENT OF GOVERNMENT

Pat Goudvis

San Francisco longshoreman loads DDT being shipped to Indonesia.

WHILE THE Export-Import Bank, the Overseas Private Investment Corporation, and even the Department of Agriculture are busy subsidizing and promoting pesticide exports, the one U.S. agency directly responsible for pesticide regulation—the Environmental Protection Agency (EPA)—is strained beyond its capacity to cope with the enormous volume of potentially dangerous chemicals under its jurisdiction. Its warehouses already contain records for over 35,000 chemicals.[1] At least 1,000 new substances are introduced every year.[2] EPA records catalogue a grisly range of toxic effects: the ability to kill, deform, mutate, and cause brain damage and cancer in living and future generations of animals, including human beings.

The EPA is hamstrung. Due to congressional funding cutbacks and industry pressure, the staff and budget employed to work on pesticides are smaller than those in any *one* of the 12 large companies which dominate the industry that the EPA is supposed to regulate.[3] The EPA has essentially thrown up its hands, allowing large numbers of "conditional" registrations of pesticides which have not yet been thoroughly tested.

The EPA's monthly enforcement reports chronicle an unending pattern of major and minor violations by pesticide companies, and illustrate the agency's inability or unwillingness to punish violations. The infrequent fines— usually less than $5,000—are insignificant to a multinational corporation. Chevron, for example, was fined only $3,200 for shipping dieldrin in violation of an EPA suspension order.[4] And a suspension order is one of the strongest actions the EPA can take. Velsicol was fined only $1,600 for failing to register, and for misbranding, the deadly phenoxy herbicide 2,4-D.[5]

✳ If domestic control seems dangerously lax, the international situation is even worse. U.S. law explicitly allows manufacturers to export banned or restricted pesticides.[6] The EPA cannot force U.S. manufacturers to cease production of any pesticide as long as it is destined for overseas use. (Other agencies could force a company to stop production if, for example, it was disabling workers at the American plant.)

➤ Moreover, companies do not have to register their pesticides. And they are free to make "for export only" any unregistered pesticide they please. They do not even have to inform the EPA of the substance's ingredients. *Unregistered pesticides now account for more than 25 percent of all U.S. pesticide exports,* according to the GAO.[7]

Unregistered Kepone—the deadly nerve toxin which seriously damaged an entire workforce and contaminated the James River in Virginia—was produced for overseas markets where it was applied to bananas grown for U.S. consumers.[8]

Since unregistered pesticides have never been independently evaluated for toxic properties, some of these chemicals may be even more dangerous than those already banned by the EPA. Completely outside of regulatory control, these toxins and their effects remain hidden behind the corporate "trade secrets" veil. Some of these substances are undoubtedly the unidentified chemical residues the FDA frequently detects in imported food. (See

Chapter Four.)

EPA notification

A S THE SECOND printing of this book went to press, the Reagan administration was preparing an all-out assault against the only regulatory impediment to unchecked pesticide dumping—a provision in FIFRA requiring "notification" of foreign governments when banned or severely restricted pesticides are exported.[9]

Within eight months of taking office, high officials of the Reagan administration had drafted a report, which was leaked to the press, recommending a virtual dismantling of the notification process. The administration's efforts on the executive level closely resembled an industry initiative in Congress led by the National Agricultural Chemicals Association (NACA). In testimony during the summer of 1981, Jack Early, president of NACA, advocated limiting the notification requirement, in a way which would render it virtually useless as a restriction on the export of hazardous pesticides.[10]

Although "notification" does not give the EPA the power to stop the export of banned or restricted pesticides, the agency developed the procedure under Congressional direction in 1979 to ensure that foreign buyers were at least *informed* about the known dangers of pesticides they were receiving.

In the past, the foreign buyer would have no indication that a pesticide was too dangerous to be registered in the United States. Under the "notification" rules, the U.S. exporter is required to send the importer a statement disclosing whether the pesticide is unregistered or whether its registration was canceled. The importer must acknowledge to the EPA that he has received this notification. The EPA then sends this acknowledgment to the State Department, which forwards the information to the appropriate official in the importing country.

The EPA is also required to tell foreign governments why it will not register a particular pesticide. The purpose of the "notification" is to help the recipient country decide for itself whether it is willing to take the risks associated with importing a particular chemical. But the pesticide industry succeeded in getting eliminated the requirement that shipments be delayed until the host government responded to the notification. The program has other critical flaws:

• The EPA does not always adequately explain to the foreign government why it restricted or banned a pesticide.

• Notifications are usually sent to local pesticide importers (such as those in Central America selling to Castle & Cooke). The pesticide importer is hardly likely to be interested in telling farmers of the product's potential dangers. Moreover, local agriculture officials are often intimately connected to the pesticide industry. "The people who use pesticides, the people who import pesticides, and the people who 'regulate' pesticides are the same people," says Dr. Harold Hubbard of the Pan American Health Organization. "It's a tight little group in each country." [11]

• The pesticide industry has ingenious ways of getting around the notification procedure. A multinational executive told Frances Miley, one of the EPA officials responsible for the new regulation, that companies will simply ship their pesticides to subsidiaries in intermediary countries, such as Switzerland, and then re-export them. [12] Switzerland, which uses almost no pesticides, will then end up with piles of useless notifications.

• A notice is required only with the *first* shipment of any particular chemical sent to a country within a calendar year. Thus the recipient government is not alerted as to *how much* of a dangerous substance it will be getting—a crucial factor in deciding whether to ban it.

• There is no enforcement mechanism. "If the companies don't do it," says Miley, "there is no penalty." [13]

In spite of these problems, the 1979 notification provisions took a significant step forward in at least beginning to regulate the flow of hazardous pesticides overseas. However, even this tentative move may not survive the anti-regulatory drive mounted by the Reagan administration.

Washington's nightmares about the pesticide time bomb

IF ANY SINGLE factor will spur the government to regulate pesticide dumping, it will be concern over its harmful effects on U.S. foreign policy. The prospect of "Made
in U.S.A." chemical time bombs ticking away around the
globe alarms the State Department. The publicity surrounding past incidents, especially the Phosvel-caused
deaths of Egyptian farmers and at least a thousand water
buffalo in 1971, and the malathion poisonings of hundreds
of Pakistanis in 1976, embarrassed Washington, to say the
least.

The State Department held a special conference in 1979
to discuss the issue, inviting representatives from Latin
America, Asia and Africa.

EPA officials unveiled their new "notification" regulations at this conference. Environmentalists led by Jacob
Scherr of the Natural Resources Defense Council argued
forcefully for an end to the unhindered export of hazardous pesticides. Corporate spokespersons such as Frederick Rarig of Rohm & Haas complained about "overlap-
ping governmental authorities . . . with myriads of regulations imposed by diverse jurisdictions." [14] Third world representatives like Samuel Gitonga of Kenya declared that
"our regulations of pesticides are in the infant stage." [15]
Delegates from several underdeveloped countries de-
scribed pesticide horror stories, including deaths, environmental destruction, and deceptive corporate sales techniques.

Everyone agreed there was a serious problem, but a
conference hosted by the State Department could not ar-
rive at a solution—that would require bucking the powerful chemical corporations. Instead the pro forma conclusion recommended "further study and better international
cooperation." Perhaps the most significant aspect of the
conference was that it took place at all. No one could now
doubt that the State Department was worried over the
chaotic export trade in hazardous pesticides.

The White House
inter-agency task force

THE WHITE HOUSE, meanwhile, inched its way toward involvement in the pesticide dumping controversy. In 1979, President Carter created an inter-agency Hazardous Substances Export Policy Task Force to prepare guidelines governing the export of hazardous goods—including pesticides, drugs, contraceptives, and Tris-treated baby clothes.

A few members of the task force—from the Council on Environmental Quality, the State Department, and the EPA—urged strong regulations to curb the surge of hazardous exports. On the other side, however, the Departments of Commerce, Agriculture, and Treasury and the Eximbank supported the pro-export positions of the chemical company lobbyists.

"The Commerce Department is leading the retreat," said Ed Cohen, White House counsel for the task force.[16] Commerce Department representatives saw the task force as an unnecessary hindrance to U.S. competitiveness in export markets. The Commerce Department's yearly pesticide export statistics obfuscate information essential to the task force. From these statistics it is impossible to distinguish between banned and approved pesticides.

Because of its deep internal divisions, the task force's final recommendations were a lopsided compromise reflecting the disagreements among the 16 executive departments and agencies involved.

The task force's major preoccupation appeared to be damage the United States may suffer if its image gets tainted with disasters linked to U.S. pesticide exports. "If the United States does not exercise special vigilance over the export of certain extremely hazardous banned or significantly restricted products which represent a substantial threat to human health or safety or the environment," the task force's final report noted, "our economic and diplomatic ties with other countries could be jeopardized. Citizens and governments of foreign countries receiving these products directly, or being adversely affected indirectly as innocent bystanders may develop increasingly hostile attitudes toward this country and its products." [17]

Specifically *excluded* from this warning and from policy recommendations, however, were the export of hazardous production facilities, or their financing by government or government-related credit institutions such as the Eximbank, OPIC, or the World Bank.

The task force also recommended more stringent notification requirements. A government would not only have to be notified before being sent a pesticide unregistered or severely restricted in the United States; it would have to make a specific *request* for the product. [18] Although an improvement over the government's "hands off" approach to hazardous pesticide exports, the task force plan was fatally flawed. Most third world countries do not have facilities for testing pesticides or evaluating their hazards. The third world governments which must make pesticide decisions are often more responsive to the needs of the multinational corporations than the welfare of their own people. Bribery, already widespread, could increase as companies tried to encourage foreign countries to purchase banned products that they could no longer sell on the home market.

In contrast to the White House task force, Ralph Nader's Health Research Group advocated automatically outlawing the export of any substance banned or restricted as too dangerous for use in the United States. A company or nation would then have to petition the EPA to prove that special circumstances in the receiving country necessitate an exception. [19]

Despite its inadequacies, the task force's recommendations were a strong symbolic step in the right direction when they were enacted into law by President Carter in his "Executive Order on Federal Policy Regarding the Export of Banned or Significantly Restricted Substances," issued five days before he left office in January 1981. Several weeks later, however, President Reagan swept away the task force's work by rescinding the executive order in its entirety. [20]

Capital Hill attempts to stop "dumping"

LATE IN 1978, Congress held hearing on the export of banned products, including pesticides, but passed no laws.

Early in 1980 and again in 1981, Rep. Michael Barnes (D-Maryland) introduced a bill which would significantly restrict the export of all hazardous products.[21] The Barnes bill would force exporters to obtain a government license before shipping their products overseas. A license would depend on these conditions being met:

• The importing country's government would have to approve and request the product.

• The U.S. government would have to determine that the potential benefits of the export outweigh the risks.

• The product would have to adhere to U.S. labeling requirements and contain instructions the importing country's people could use.

• Perhaps more significant, ingredients of banned products could not be exported to formulate those banned products overseas.

Barnes held committee hearings on his bill in mid-1980, but passage seems unlikely, given the strong pro-export mood of the 96th Congress. Nevertheless, the Barnes bill seems to lay the legislative groundwork for regulation of hazardous exports by the U.S. Congress—regulation that will only be possible when growing citizen activism challenges the chemical industry.

CHAPTER NINE

BREAKING THE CIRCLE OF POISON

Pesticide dust is stirred by tractor.

Rick Meyers/LA Times

S TOP USING US as a dumping ground!" exclaimed Kenya's minister for water development, Dr. Gikonyo Kiano, at a U.N. Environment Program meeting. "Kenya detests the use of developing countries as experimental or dumping grounds for chemical products that have been banned or have not been adequately tested."[1]

Throughout the third world, activists like Kiano are beginning to resist chemical colonialism. They are organizing to resist the pesticide onslaught, even though dissenting can be dangerous under the repressive governments in many third world countries.

The Philippines

T HE FARMER'S Assistance Board was organized several years ago by Filipino peasants and students to study pesticides. The group places the blame for the huge volume of pesticides used in Philippine agriculture squarely on big export producers such as Castle & Cooke and Del Monte as well as the promotional research of the International Rice Research Institute. The companies demand the highest yields and blemish-free products. Both demands contribute to dependence on pesticides.

The Farmer's Assistance Board treads a delicate, dangerous line under Marcos' martial law regime, but its description of the impact of pesticides on Philippine farms is revealing: "It may well be that the gains of modern agricultural technology are wiped out by the destruction of the material bases for food production: the land, the air, and the water, and the living beings that derive their sustenance and survival from these elements."[2]

Malaysia

IN NEARBY Malaysia, pesticide residues on local food crops have galvanized local consumers into action. The Consumers Association of Penang has discovered organochlorine pesticides such as DDT, aldrin, BHC, dieldrin, and chlordane—all banned in the United States—in Malaysia's rainwater, soil, drinking water, and food crops.

The Consumers Association is the largest and loudest citizens organization in the third world focusing specifically on consumer rights. For 10 years the association has been monitoring industrial pollution, deforestation, war toys, flammable teddy bears, adulterated foods, and pesticide poisoning of workers and residues on food in Malaysia. It collected evidence of nearly 100 pesticide poisonings in 1975, mostly from malathion and paraquat. Heavy pesticide use is destroying the nation's fish supply, the Consumers Association points out. Dramatic decreases in the Muda River catch due to endosulfan, BHC, and malathion poisoning have led to "severe economic hardship and nutritional deficiencies among the poorer paid farmers."[3] The association has been pressuring the Malaysian government to tighten its regulations on pesticides, with some recent success.

The United Nations

THESE EXAMPLES from the third world—along with parallel efforts in the industrial countries—represent the beginning of a worldwide attempt to regulate the flow of toxic pesticides. A symbolic step took place in December 1979, when the General Assembly of the United Nations passed a resolution urging member states to exchange information on hazardous chemicals (and pharmaceuticals) that have been banned in their countries, and to discourage exporting those products to other countries.

"But simple notification is not good enough," says Noel Brown, North American representative for the U.N. Environment Program. Brown, a Jamaican, is one of the U.N.'s most articulate voices against the use of underdeveloped countries as dumping grounds for the industrialized world. "We would like to have more than simply telling a government a substance is banned," he continues. "Why is it banned? We want basic chemical information. The side effects: what dangers are there? We want full disclosure."[4]

Brown is trying to bring the issue out in the open. The U.N. has little official power to prevent the sales of hazardous materials, he says, but focusing international attention on the sellers and producers of the materials may at least make "dumpers" think twice.

Having seen his native Caribbean Sea heavily damaged by agricultural runoff, Brown is particularly concerned about pesticides. He is encouraging local doctors to develop labels for poisonous pesticides which use special color codes or graphics specifically oriented to the local population, with warnings in local languages.

Since there is no reliable source of information about the global movement of toxic materials, Brown would like to expand the U.N.'s International Registry of Potentially Toxic Chemicals so it could track them from the manufacturers—mostly in the industrial countries—to their destinations in the third world.

In the summer of 1980, the Organization for Economic Cooperation and Development (OECD) asked all the industrial countries to observe the U.S. notification procedures in order to standardize the global trade in toxic chemicals. Based in Paris, the organization is just beginning a comprehensive effort to compile data on toxic chemicals and their harmful effects. It would be the first central collection of data on chemical poisons used around the world.

Breaking the circle of poison

WHILE INVESTIGATING pesticide dumping we have relearned a hard fact: where a few executives from a handful of multinational corporations and their government allies are allowed to make decisions affecting entire peoples, "business as usual" will not serve the majority. The problem is not simply unethical corporate executives; the solution is not simply exchanging them for more compassionate, socially responsible types. No, we have concluded that unless all a society's important decisions—including economic decisions like the development and marketing of agricultural chemicals—are made more democratically, the majority will suffer. By "more democratically" we mean that all the people affected—in this case, factory workers, farm workers, consumers—should take part in making the key decisions.

Yet our society is moving in exactly the opposite direction—toward more and more concentration of economic power in the hands of a few corporations.

We have no blueprint for achieving this goal of greater participation. But we are confident that more and more Americans are realizing—whether the issue is food, energy or health policy—that the answer is not to make the powerful more responsible, but to redistribute the power. This realization challenges each of us. It means that our responsibility is not merely to express outrage to our "leaders," although that is clearly the beginning. Each of us must figure out how to begin to take greater responsibility for the economic system. The first step may well be learning much more about how economic decisions are made *now*, and then examining our roles as workers, consumers, parents, citizens. In each case we must ask if our actions serve to reinforce the highly anti-democratic concentration of economic power, or if our actions begin to challenge it.

For us, this investigation has contributed to a second important realization. *Circle of Poison* reveals that the problem cannot be reduced to that of the "rich countries" exploiting the "poor countries." Such a formulation turns the issue into "them versus us." But the majority in *both* the industrial nations and the third world are victims of the circle. When we understood this, we began to understand our ties with third world people in a new way. The differences in our material standards of living too often obscure our similarities—a common powerlessness in facing the increasing concentration of private power in the hands of a relatively few global companies. The reality of global corporate power, here reflected in the pesticide trade, forces us to seek solutions involving new ways of working *with* third world people for a worldwide redistribution of economic power. We must begin to see third world people not as a burden or a threat, but as allies.

Momentum for change

B UT WHERE TO BEGIN? Here are our suggestions as to how you can help build the momentum for change. First, your letters can be ammunition. They can inform those in power that their actions are being scrutinized—that we are not blind to the dangers of pesticide proliferation. Second, your letters can help concerned people *within* the government by demonstrating that there *is* widespread concern.

We suggest you write letters. (The addresses are in Appendix A.) Write to your Congressperson, recommending support for the Barnes bill (H.R. 2439) which would restrict pesticide dumping. Also write to Congressman George Brown Jr. (D-California) and Jonathan Bingham (D-New York) to request that their House committees hold hearings on the dumping of pesticides and other dangerous products. Write to Noel Brown of the U.N. Environment Program, stating what you think needs to be done to end pesticide dumping. And write to corporate dumpers themselves, asking them to respond to the damning evidence in this book.

send a copy of *Circle of Poison* along with your letter.

In addition, you can help expose this scandal to the public. Why not try to interest your local newspaper in a story based on this investigation? Many columnists and editorial writers are looking for controversial issues. Pesticide dumping is also a good radio talk show topic. Why not suggest it?

Take this book to your or your child's classroom, to your church or study group, or to your union meeting. Since it affects all of us, it provokes discussion. Most important, it is educational, helping "demystify" economics. By learning about the proliferation of hazardous pesticides, we can learn about how our economic system works. Ask your local library to order and display it.

We also encourage those of you who may be traveling abroad to help push forward this research. You don't have to be an experienced investigator. You just have to keep your eyes open and be willing to ask questions. Much of the research for this book relied on such first-hand accounts—on pesticide labels collected, photos taken by friends, and on interviews with local farmers and pesticide promoters. Please send us what you turn up.

If this issue is one that ignites you into action, work with other people. Contact one of the action groups listed in Appendix A. See what you can do to support or extend their initiatives. We have included some third world groups there, too. They need information that we, living here, can supply about dumping practices of U.S.-based corporations.

As for us, our hope is that this book will help transform the issue from one receiving sporadic press coverage and scattered activist attention, to one that spurs an international movement. That means getting the word out. We need your financial help to be able to send out books to policy and opinion makers both here and in the third world, to mail press releases, and to place ads. If we can secure the financial support, we would also like to assemble a small group of activists from both the industrial and third world countries for a working session in 1981 to map out a more coordinated strategy for information sharing, media coverage and government pressure.

This book is but a beginning. It will only be of value if your actions give it the power to help break the pesticide circle of poison.

BUREAUCRACY GLOSSARY

AID   U.S. Agency for International Development (part of State Department)

EPA   U.S. Environmental Protection Agency

FAO   Food and Agriculture Organization (part of UN)

FDA   U.S. Food and Drug Administration (part of Department of Human Health Services—HHS—formerly HEW)

FIFRA   Federal Insecticide, Fungicide, and Rodenticide Act (the basic legislation governing pesticides)

GAO   General Accounting Office (investigative arm of U.S. Congress)

ICAITI   Instituto Centroamericano de Investigacion y Technologia Industrial (nonprofit research organization funded by UN and 5 Central American countries)

ICD   Industry Council for Development (the new ICP)

ICP   Industry Cooperative Program (a vehicle created by multinational corporations to influence the UN)

IICA   Interamerican Institute for Agricultural Sciences (abbreviation from Spanish) (technical support organization in Latin America)

ILO   International Labor Organization (part of UN)

IRRI   International Rice Research Institute (in the Philippines, part of the Green Revolution research network)

OPIC   Overseas Private Insurance Corporation (U.S. agency for insuring company investments overseas against threat of nationalization, expropriation, etc.)

OSHA   Occupational Safety and Health Administration (part of Department of Labor)

PAHO   Pan American Health Organization (component of World Health Organization)

UN   United Nations

UNDP   United Nations Development Program

UNEP   United Nations Environment Program

USDA   United States Department of Agriculture

WHO   World Health Organization (part of UN)

APPENDIX A

Your letters can help stop the dumping of hazardous pesticides, as we suggest in Chapter IX. Here are the addresses:

Congressmen active or potentially active on the dumping issue:
    Rep. Jonathan Bingham (D-NY)
    2262 Rayburn House Office Bldg.
    Washington, DC 20515

    Rep. George Brown Jr. (D-CA)
    2342 Rayburn House Office Bldg.
    Washington, DC 20515

    Consumers Union of U.S., Inc.
    256 Washington St.
    Mt. Vernon, NY 10550

International Groups:
    International Organization of Consumers Unions (IOCU)
    Emmastraat 9
    N-2595 EG
    The Hague, Netherlands

United Nations:
    Noel Brown
    United Nations Environment Programme
    New York Liaison Office
    Room A-3608
    United Nations Plaza
    New York, NY 10017

Corporate pesticide dumpers:
    Ortho (Chevron Chemical Company)
    274 Brannan Street
    San Francisco, CA 94107

    Monsanto Company
    800 North Lindbergh Blvd.
    St. Louis, MO 63166

    ICI America's, Inc.
    Agricultural Chemicals Division
    P.O. Box 208
    Goldsboro, NC 27530

Castle and Cooke
50 California Street
San Francisco, CA 94119

Velsicol Chemical Corporation
341 East Ohio Street
Chicago, IL 60611

Amvac Chemical Corporation
4100 East Washington Blvd.
Los Angeles, CA 90023

In the United States a number of public-interest groups are focusing their attention on the abuse of pesticides in this country and around the world. The Natural Resources Defense Council, for example, initiated the campaign against the FIFRA export loophole ten years ago. All of the following groups are active, can provide information on pesticides, and welcome help to continue their struggle.

Natural Resources Defense Council
1725 I Street, NW
Washington, D.C. 20006

Environmental Defense Fund
1525 18th Street, NW
Washington, D.C. 20036

California Rural Legal Assistance
1900 K Street
Sacramento, California 95814

Coordinating Committee on Pesticides
1057 Solano Avenue
No. 106
Albany, California 94706

Pesticides Unit
Cal-OSHA
455 Golden Gate Avenue
San Francisco, California

(California's state occupational safety and health unit is one of the few government agencies seriously attempting to regulate pesticide production and use.)

United Nations Environment Programme
New York Liaison Office
Room A-3608
United Nations Plaza, New York 10017
(UNEP has information on environmental issues around the world.)

National Association of Farmworkers Organizations
1332 New York Avenue, N.W.
Washington, D.C. 20005

Third World Groups.
   Organized resistance to pesticides in the third world is scattered, and dependent often on the level of repression faced by dissidents in the country. The organizations listed below have emerged in the forefront of the fight against pesticides in the third world.

Consumers Association of Penang
27 Kelawei Road
Pulau Pinang,
Malaysia

Friends of the Earth
7 Cantonment Road
Penang,
Malaysia

Farmer's Assistance Board
P.O. Box AC-623
Quezon City,
Philippines

Environment Liaison Centre
P.O. Box 72461
Nairobi,
Kenya

APPENDIX B
Costa Rica—1978 Imports of Selected Pesticides

Pesticide	Exporter(s)	Importer(s)
Aldrin/Dieldrin	Shell	Ortho
DDT	Stauffer	Ortho
BHC	Hooker	Ortho
Lindane	Hooker	—
DBCP	Dow	Standard Fruit (Castle & Cooke)
Heptachlor/ Chlordane	Velsicol	Ortho, Velsicol
Endrin	Velsicol	Ortho, Velsicol
Phosvel	Velsicol	Velsicol
Toxaphene	Hercules	Ortho/ Rohm & Haas
Malathion	FMC, Am. Cyanamid, Chevron	Ortho
Paraquat	Imperial Chemicals (ICI)	Standard Fruit (Castle & Cooke)
Parathion	Kerr-McGee, Bayer	Ortho
Silvex	Dow	Rohm & Haas
2,4-D	Dow, BASF	—
2,4,5-T	BASF	—

Source: Ministry of Agriculture Government of Costa Rica

APPENDIX C
Colombia, Major Importers of Selected Pesticides (1979)

Company	Pesticides
American Cyanamid	parathion, malathion
Bayer	DDT, toxaphene, parathion
BASF	2,4,-D
Celamerck	aldrin, dieldrin, endrin, heptachlor, chlordane, parathion, toxaphene, lindane, DDT, 2,4,-D
Ciba-Geigy	2,4-D
Dow	2,4,-D
Dupont	EPN, parathion
Hoechst	parathion, malathion, DDT
Monsanto	parathion
Proficol	aldrin, malathion, parathion, lindane, BHC, heptachlor, chlordane, DDT, dieldrin, EPN, endrin
Schering	aldrin, heptachlor, BHC, parathion
Shell	aldrin, dieldrin, DDT, endrin, 2,4-D, parathion
Union Carbide	DDT, parathion, EPN, heptachlor, chlordane, endrin
Velsicol	heptachlor, chlordane, endrin, EPN

Source: Ministry of Agriculture
Government of Colombia

TABLE ONE
Selected List of Chemical Companies Producing, Buying, and/or Selling Hazardous Pesticides in the Third World

COMPANY (U.S. unless otherwise noted)	PESTICIDES		
	A Banned or Heavily Restricted	B Under Review	C Unrestricted
Allied Chemical	Kepone, Mirex	—	—
Amvac	DBCP	—	—
American Cyanamid	Kepone, Mirex	Toxaphene, 2,4-D	Malathion Parathion
BASF (W. Germ.)	2,4,5-T	2,4-D	—
Bayer (W. Germ.)	DDT, Heptachlor, Lindane	Toxaphene	Parathion
Celamerck (W. Germ.)	Aldrin, Dieldrin, DDT, Endrin, Heptachlor, Chlordane, Lindane. 2,4,5-T	Toxaphene, 2,4-D	Parathion
Chevron	DDT, Aldrin, Dieldrin, Heptachlor, Chlordane, Endrin, Lindane, BHC, Silvex	Toxaphene, Paraquat	Malathion
Ciba-Geigy (Swiss)	—	2,4-D	—
Dow	2,4,5-T, Silvex, DBCP	2,4-D	—
Dupont	—	EPN	Parathion
FMC	Heptachlor	—	Malathion

| COMPANY (U.S. unless otherwise noted) | PESTICIDES | | |
	A Banned or Heavily Restricted	B Under Review	C Unrestricted
W.R. Grace	—	Toxaphene	—
Hercules	—	Toxaphene	—
Hoechst (W. Germ.)	DDT	—	Parathion Malathion
Hooker	BHC, Lindane, Mirex	—	—
Imperial Chemicals (UK)	BHC, Aldrin	Paraquat	—
Kerr-McGee Chem.	—	—	Parathion
Monsanto	—	2,4-D	Parathion
Montrose	DDT, Endrin	—	Parathion
Nissan (Jap.)	—	EPN	—
Pfizer	—	—	Malathion
Rohm & Haas	Silvex	Toxaphene	Parathion
Schering (W. Germ.)	Aldrin, BHC, Heptachlor	—	Parathion
Shell (UK-Neth.)	Aldrin, Dieldrin, DDT, DBCP, Endrin, 2,4,5-T	2,4-D	Parathion
Stauffer	DDT, Dieldrin	EPN,2,4-D	Malathion, Parathion
Sumitomo (Japan)	—	—	Malathion
Union Carbide	DDT, Mirex, Heptachlor, Chlordane, Endrin	EPN	Parathion
Velsicol	Chlordane, Heptachlor, Phosvel, Endrin	EPN	Parathion

COMPANY	PESTICIDES		
(U.S. unless	A	B	C
otherwise noted)	Banned or	Under Review	Unrestricted
	Heavily		
	Restricted		

Note: **Category A:** Those which are banned or heavily restricted inside the U.S. Most uses for these products have been outlawed, but important uses for some remain, such as termite control for Chlordane. Certain pesticides have recently been discontinued, including DBCP (Dow) and Kepone (Allied Chemical), but are included because they were important products for the companies involved.

Category B: Those which are under review for future regulatory action. Toxaphene, Paraquat and EPN are termed "suspect chemicals" by EPA.

Category C: Those which are unrestricted in the U.S. but which have caused human deaths in the third world. Parathion is reportedly the number one killer among all hazardous pesticides, banned or not.

Sources: This table was compiled from a variety of official and unofficial sources, including the EPA publication "Suspended and Cancelled Pesticides"; EPA production reports; company records and advertisements, personal interviews and observations, and government import statistics for seven third world countries.

TABLE TWO
Pesticides Used in Foreign Countries on Food Exported to the United States

		Number of pesticides		
Commodity	*Countries surveyed*	*Allowed, recommended or used in the U.S.*	*Any residue prohibited (no U.S. tolerance)*	*Not detectable with FDA tests*
Bananas	Colombia, Costa Rica Ecuador, Guatemala, Mexico	45	25	37
Coffee	Brazil, Colombia, Costa Rica, Ecuador, Guatemala, Mexico	94	76	64
Sugar	Brazil, Colombia, Costa Rica, Ecuador, Guatemala, India, Thailand	61	34	33
Tomatoes	Mexico, Spain	53	21	28
Tea	India, Sri Lanka	24	20	11
Cacao	Costa Rica, Ecuador	14	7	7
Tapioca	Thailand	4	4	1
Strawberries	Mexico	13	—	5
Peppers	Mexico	12	—	4
Olives	Italy, Spain	20	14	8
Totals		340	201 (59%)	198 (58%)

Source: GAO

TABLE THREE
Pesticides in Imported Coffee Beans
(1974-1977)

Country of Origin	No. of Samples	No. with Residues
Angola	1	1
Brazil	2	2
Colombia	21	5
Costa Rica	2	0
Dominican Republic	1	0
Ecuador	10	6
El Salvador	2	1
Guatemala	5	2
Haiti	1	1
Honduras	2	1
India	4	4
Indonesia	1	1
Ivory Coast	2	1
Kenya	1	0
Mexico	5	4
New Guinea	2	1
Nicaragua	2	0
Panama	1	0
Peru	5	2
Rwanda	1	1
Uganda	1	1
Venezuela	2	1
Total (22)	74	35

Percentage Contaminated: 47.3%
Pesticides Detected: DDT, DDE, BHC, Lindane, Dieldrin, Hepta-
chlor, Diazinon, Malathion
Source: FDA

TABLE FOUR
Pesticide manufacturers, formulators and importers in the Philippines

Categories: A Manufacture, formulate, market B Formulate and market C Formulate only D Market only

	Category	Origin	Companies with Formulation Plant
1. Agchem Manufacturing Corp.	A	Filipino	X
2. BASF (Phils.), Inc.	D	Germany	
3. Bayer Philippines, Inc.	B	Germany	X
4. Ciba-Geigy Agrochemicals	C	Switzerland	X
5. Cyanamid Philippines, Inc.	B	USA	a
6. Dow Chemical International	D	USA	
7. Du Pont Far East	B	USA	X
8. Eli Lilly Philippines, Inc.	B	USA	a
9. Hoechst Philippines, Inc.	B	Germany	X
10. Interchem Philippines	B	Filipino	a
11. Macondray Company, Inc.	B	USA	a
12. Marsman & Company, Inc.	B	USA	X
13. Monsanto Philippines, Inc.	D	USA	
14. Pfizer Incorporated	D	USA	X
15. Planters Products, Inc.	D	Filipino	X
16. Rohm & Haas Philippines, Inc.	D	USA	
17. Shell Chemical Co. (Phils.)	B	Filipino	X
18. Transworld Trading Co. Inc.	D	Filipino	
19. Union Carbide Phils., Inc.	B	USA	X

	Category	Origin	Companies with Formulation Plant
20. Velsicol Chemical Company	D	USA	
21. Warner Barnes and Company	B	Filipino	X
22. Zuellig Agro-Chemicals	B	Filipino	a

Source: Fertilizer and Pesticide Authority Ministry of Agriculture, The Philippines

a Other companies formulate for them

TABLE FIVE
Recent North American Seed Company Acquisitions

NEW OWNER	SEED COMPANY
Anderson Clayton	Paymaster Farms
	Tomaco-Genetic Giant
Cargill	Dorman Seeds
	Kroeker Seeds
	PAG
Celanese	Cepril Inc.
	Moran Seeds
	Harris Seeds
Central Soya	O's Gold Seed Co.
Ciba-Geigy	Funk Seeds Intern'l.
	Stewart Seeds
Diamond Shamrock	Taylor-Evans Seed Co.
FMC	Seed Research Association
Garden Products	Gurney Seeds
Hilleshoeg/Cardo	Intern'l. Forest Seeds Co.
Intern'l. Multifoods	Baird Inc.
	Lynk Bros.
ITT	O.M. Scott & Sons
	Burpee Seeds
Kent Feed Co.	Teweles Seed Co.
KWS AG	Coker
Monsanto	Farmers' Hybrid Co.
NAPB (Olin & Royal Dutch/ Shell)	Agripro, Inc.
Pioneer Hi-Bred	Lankhart
	Lockett
	Arnold Thomas Seed Co.
	Petersons

NEW OWNER	SEED COMPANY
Pfizer	Clemens Seed Farms
	Jordan Wholesale Co.
	Trojan Seed Co.
	Warwick Seeds
Purex	Advanced Seeds
	Ferry-Morse Seeds
	Hulting Hybrids
Sandoz	National-NK
	Northrup-King
	Rogers Brothers
Southwide, Inc.	Delta Pineland
Tate & Lyle	Berger & Plate
Tejon Ranch Co.	Waterman-Loomis Co.
Union Carbide	Keystone Seed Co.
	Jacques Seeds
	Amchem Products
Upjohn	Asgrow Seeds

Source: International Coalition for Development Action

NOTES

CHAPTER ONE

1. Proceedings of the U.S. Strategy Conference on Pesticide Management, U.S. State Rept., June 7-8, 1979, p. 33.
2. "Report on Export of Products Banned by U.S. Regulatory Agencies," H. Rept. No. 95-1686, Oct. 4, 1978, p. 28.
3. Ibid.
4. "Better Regulation of Pesticide Exports and Pesticide Residues in Imported Foods Is Essential," U.S. GAO, Rept. No. CED-79-43, June 22, 1979, pp. iii, 39.
5. FIFRA, 7 U.S.C. 1360.
6. "New Pesticides Must Now Be Economic Winners," *Chemical Age*, Feb. 17, 1978; Dr. Jay Young, Chemical Manufacturers Association, telephone interview with authors, Oct. 1979.
7. President's Hazardous Substances Export Policy Working Group, Fourth Draft Report, Jan. 7, 1980, p. 6.
8. Dr. Hal Hubbard, telephone interview with authors, June 1, 1977.
9. "Importacion de Pesticides," Ministerio de Agricultura y Gamaderia, Costa Rica, 1978.
10. "Listudo de Pesticides Registrados en el Departamento de Sanidad Vegetal," Ministry of Agriculture, Ecuador.
11. "Report on Plant Protection in Major Food Crops in Malaysia," FAO, Dec. 1977, pg. 59.
12. "Plaguicidas de Uso Agricola, Défoliantes y Reguladores Fisiologicos de las Plantas Registrados en Colombia," Ministerio de Agricultura, Colombia, June 30, 1979.
13. "Erosion and Soil Depletion-By-Products of Castle & Cooke Operation," Communications, M.S.P.C., April 1978.
14. Thomas O'Toole, "Over 40 Percent of World's Food Is Lost to Pests," *Washington Post*, March 6, 1977.
15. Douglas Starr, " 'Pesticide Poisoning Alarming,' says FAO," *Christian Science Monitor*, Feb. 1, 1978.
16. Ibid.; and Lappé and Collins, op. cit., p. 64.
17. "Better Regulation of Pesticide Exports and Pesticide Residues in Imported Food Is Essential," U.S. GAO Rept. No. CED-79-43, June 22, 1979, p. 1.
18. Jeanie Ayres, "Pesticide Industry Overview," *Chemical Economics Newsletter*, Jan.-Feb. 1978, p. 1.
19. Lappé and Collins, op. cit., p. 41.
20. "An Environmental and Economic Study of the Consequences of Pesticide Use in Central American Cotton Production," Final Report, Instituto Centro-Americano de Investigacion y Technologia Industrial (I.C.A.I.T.I.), Jan. 1977, pp. 149, 155, 161.
21. Lappé and Collins, op. cit., p. 67.
22. Osawa Yasuo, "Banana Plantation Workers Strike in the Philippines," *New Asia News*, May 1980, p. 7.

23. Dr. Lou Falcon, telephone interview with authors, May 21, 1979.
24. *Agriculture: Toward* 2000, U.N. Food and Agriculture Organization (FAO), Rome, July 1979, p. 82.
25. 5 U.S.C. §552, Freedom of Information Act.

CHAPTER TWO

1. Douglas Starr, " 'Pesticide Poisoning Alarming,' says FAO," *Christian Science Monitor*, Feb. 1, 1978.
2. Dr. V. H. Freed, personal interview with authors, January 4, 1980.
3. Laurie Becklund and Ronald Taylor, "Pesticide Use in Mexico—A Grim Harvest," *Los Angeles Times*, April 27, 1980.
4. Ibid.
5. Ibid.
6. "An Environmental and Economic Study of the Consequences of Pesticide Use in Central American Cotton Production," Final Report, Instituto Centro-Americano de Investigacion y Technologia Industrial (I.C.A.I.T.I), Jan. 1977, pp. 97 and 98.
7. Ibid., p. 195.
8. Ibid., p. 2.
9. Alan Riding, "Guatemala: State of Siege," *New York Times Magazine*, August 24, 1980, p. 20.
10. I.C.A.I.T.I. Report, p. 164.
11. Dr. H. L. Falk, telephone interview with authors, May 25, 1979.
12. I.C.A.I.T.I. Report, pp. 128-32.
13. Riding, op. cit., p. 20.
14. Ibid.
15. Edward Baker, M.D., et al., "Malathion Intoxication in Spray Workers in the Pakistan Malaria Control Program," U.S. Agency for International Development, pp. 3, 7.
16. I.C.A.I.T.I. Report, p. 29.
17. Robert Chambers, GAO, personal interview with authors, March 17, 1980.
18. Lucas Brader, personal interview with authors, June 7, 1979.
19. Dr. Fred Whittemore, personal interview with authors, March 17, 1980.
20. U.S. Strategy Conference on Pesticide Management, U.S. State Department, workshop discussions, June 7-8, 1979.
21. Whittemore interview, March 17, 1980.
22. Carta Mensual, "Boletiṇ Informativo de la Division de Supervision de Insumos Agricolas, Ministerio de Agricultura, Government of Colombia, Oct. 1979, pp. 2-3.
23. Whittemore interview, March 17, 1980.
24. Freed interview, Jan. 4, 1980.

CHAPTER THREE

1. "Notice of Intent to Cancel the Registrations of Pesticide Products Containing Dibromochloroprophane (DBCP), and Statement of Reasons," U.S. Environmental Protection Agency, signed by Douglas Castle, administrator, Oct. 29, 1979.
2. Confidential interview with authors.
3. Ibid.
4. Ibid.
5. Form 10-K, American Vanguard Corp. (Amvac), Dec. 31, 1977, U.S. Securities and Exchange Commission.
6. Confidential interview with authors.
7. Form 10-K, Amvac.
8. *Dow* v. *Durham*, U.S. District Court No. 69-2162-DWW, Central District Court of California.
9. Frederick C. Klein, "Small Chemical Firm Has Massive Problems with Toxic Products," *Wall Street Journal*, Feb. 13, 1978, p. 1.
10. Alpern, Gram and Bishop, "The Phosvel Zombies," *Newsweek*, Dec. 13, 1976, p. 38.
11. Robert Chambers, GAO, personal interview with authors, June 4, 1979.
12. Alberto Donadio, letter to authors, June 8, 1979; and Elkin Bustamante, Ministerio de Agricultura, interview with authors, June 7, 1979.
13. Peter Milius and Dan Morgan, "Hazardous Pesticides Sent as Aid," *Washington Post*, Dec. 8, 1976, p. 1.
14. Keven Shea, "Nerve Damage," *Environment*, Nov. 1974, pp. 6-10.
15. "Importacion de Pesticidas," Ministerio de Agricultura y Ganaderia, Costa Rica, 1978.
16. "Basic Supply and Marketing Data for Agro-Pesticides in Indonesia," ARSAP/Pesticides, FAO (Bangkok), Jan. 1980, pp. 30-31.
17. Alberto Donadio, letter to authors, July 20, 1977; and Daniel Samper, letter to authors, Feb. 25, 1980.
18. Alberto Donadio, letter to authors, June 8, 1979; and Elkin Bustamante, Ministerio de Agricultura, interview with authors, June 7, 1979.
19. Ricardo Deang, personal interview with authors, June 7, 1979.
20. Fernando Mazariegos, personal interview with authors, June 8, 1979.
21. Richard Blewitt, telephone interview with Terry Jacobs, Center for Investigative Reporting, July 31, 1979.
22. Peter Milius and Dan Morgan, "EPA Challenging Four Velsicol Pesticides," *Washington Post*, Dec. 14, 1976, p. 1.
23. Federal Register, March 24, 1978, p. 12374.
24. Ruby Compton, Esq., and Faith Thompson Campbell, Ph.D., testimony before Subcommittee on Dept. Investigation, Oversight and Research of the House Committee on Agriculture, April 26, 1977.
25. "Importacion de Pesticidas," Ministerio de Agricultura y Ganaderia, Costa Rica, 1978.

26. "Plagucidas de Uso Agricola, Defoliantes y Reguladores Fisiologicos de las Plantas Registrados en Colombia," Ministerio de Agricultura, Colombia, June 30, 1979, p. 14.
27. U.S. embassies in Lima and Guatemala, unclassified telegrams to the U.S. Dept. of Agriculture, May 17, 1977.

CHAPTER FOUR

1. Report on Export of Products Banned by U.S. Regulatory Agencies, U.S. House Report No. 95-1686, Oct. 4, 1978, p. 28.
2. "Federal Efforts to Regulate Pesticide Residues in Food," Statement of Henry Eschaege, GAO, before Subcomm. on Oversight and Investigations, House Comm. on Interstate and Foreign Commerce, Feb. 14, 1978, p. 12.
3. "Pesticides in Mexican Produce (FY '79)," Chapter 5, FDA Compliance Program Guidance Manual, TN-78-194, Dec. 15, 1978, Attachment H, p. 1.
4. "Pesticides in Imported Coffee Beans (July 1974-May 1977 and August 1977-Oct. 1978)," FDA Compliance Program Evaluation, report in "U.S. Export of Banned Products," hearings, July 11-13, 1978, pp. 210-11.
5. "Pesticide Contamination of Imported Flowers," *Morbidity and Mortality Weekly Report*, HEW Pub. No. CDC-77-8017, April 29, 1977, p. 143.
6. "Foreign Meat Inspection 1978," U.S. Dept. of Agriculture, March 1979.
7. William Mitchell, "Study Finds Meat Can Be Dangerous to Your Health," Knight News Service report in *San Francisco Examiner*, May 1, 1979, p. 8.
8. "Illegal Pesticides Found in Imported El Salvador Meat," Associated Press report in unknown newspaper, April 1980.
9. Richard Mikita, U.S. Dept. of Agriculture, personal interview with authors, March 18, 1980.
10. "Better Regulation of Pesticide Exports and Pesticide Residues in Imported Foods Is Essential," U.S. GAO, Rept. No. CED-79-43, June 22, 1979, pp. iii, 39.
11. Ibid.

CHAPTER FIVE

1. Richard Blewitt, vice-president of public relations, telephone interview with Terry Jacobs, Center for Investigative Reporting, July 31, 1979.
2. "Proceedings of the U.S. Strategy Conference on Pesticide Management," U.S. State Dept., June 7, 1979, p. 30.
3. There are no statistics that could provide a precise estimate. This rough estimate is drawn from impressions we have received from reports and conversations with government and corporate officials and others with knowledge of third world agriculture. In their book *Food First*, Lappé and Collins refer to an estimate by the chief of the

Plant Protection Service of the U.N.'s Food and Agriculture Service, W.R. Furtick: the "vast majority" of pesticide use in the third world is on export crops.

4. "An Environmental and Economic Study of the Consequences of Pesticide Use in Central American Cotton Production," Final Report, Instituto Centro-Americano de Investigacion y Technologia Industrial (I.C.A.I.T.I.), Jan. 1977, p. 26.
5. Thomas O'Toole, "Over 40 Percent of World's Food Is Lost to Pests," *Washington Post*, March 6, 1977.
6. Ibid.
7. Frances Moore Lappé and Joseph Collins, *Food First: Beyond the Myth of Scarcity* (New York: Ballantine, 1979), p. 289.
8. "Basic Supply and Marketing Data for Agro-Pesticides in Indonesia," ARSAP/Pesticides, FAO (Bangkok), Jan. 1980, pp. 15-16.
9. Eleanor McCallie and Frances Moore Lappé, *The Banana Industry in the Philippines: An Informal Report,*" (San Francisco: Inst. for Food and Develop. Policy, 1977), p. 8.
10. David Pimentel, et al., "Pesticides, Insects in Foods, and Cosmetic Standards," *Bio-Science*, March 1977.
11. Overseas Private Investment Corporation, Annual Report, 1973.
12. McCallie and Lappé, op. cit., p. 1.
13. Fr. Jerome McKenna, telephone interview with authors, April 7, 1980.
14. Ibid.
15. Larry Rich, "Castle & Cooke, Inc.," report prepared for Big Business Day campaign, March 1980, p. 15.
16. McKenna interview, April 7, 1980.
17. Ibid.
18. Ibid.
19. Lappé and Collins, op. cit, p. 227.
20. Peter Dorner, "Export Agriculture and Economic Development," Land Tenure Center, University of Wisconsin, Madison, statement before the Interfaith Center on Corporate Responsibility, New York, Sept. 14, 1976, p. 6.
21. National Academy of Sciences, Committee on Genetic Vulnerability of Major Crops, *Genetic Vulnerability of Major Crops*, report, Washington, D.C., 1972.
22. Benton Rhoades, personal interview with authors, March 21, 1980.
23. "Basic Supply and Marketing Data for Agro-Pesticides in the Philippines," ARSAP/Pesticides, FAO (Bangkok), Feb. 1980, p. 30.
24. Between 1961-65 and 1976, rice production in the Philippines increased by two-thirds. Andrew Pearse, "A Case for Peasant-Based Strategies," United Nations Research Institute for Social Development, Report No. 79.1, Geneva, May 1979, p. 36. And rice production has continued to increase, according to the FAO *Production Yearbook*, 1978.
25. Ho Kwon Ping, "The Mortgaged New Society," *Far Eastern Economic Review*, June 29, 1979, citing Asian Development Bank and World Health Organization reports.

26. Lappé and Collins, op. cit., p. 146.
27. "Third World Seen Losing War on Rural Poverty," *ILO* (International Labor Office) *Information*, Feb. 1980, p. 8.
28. Ibid.
29. Kenneth Bachman and Leonardo Paulino, "Rapid Food Production Growth in Selected Developing Countries: A Comparative Analysis of Underlying Trends 1961-76," Research Report 11, October 1979 (Washington: International Food Policy Research Institute), p. 29.
30. Ibid.

CHAPTER SIX

1. Dr. Lou Falcon, telephone interview with authors, May 21, 1979.
2. Michael Moran, personal interview with authors, March 12, 1980.
3. René Montmeyor, telephone interview with authors, Feb. 20, 1980.
4. Julie Kosterlitz, letter to authors, May 11, 1979.
5. Dr. Harold Hubbard, telephone interview with authors, June 1, 1979.
6. Frank Penna, telephone interview with authors, March 25, 1980.
7. Julie Kosterlitz letter, May 11, 1979.
8. Dr. Fred Whittemore, personal interview with authors, March 17, 1980.
9. Frank Penna interview, March 25, 1980.
10. Ibid.
11. Gary Jones, Dow Chemical Corp., telephone interview with authors, June 11, 1979.
12. P.R. Mooney, *Seeds of the Earth* (London: International Coalition for Develop. Action, 1979), p. 117.
13. "List of Standardized Pesticides as Revised by the Pesticide Review Committee on June 16, 1979," provided by G.A. Patel.
14. G.A. Patel, India Institute of Management (Ahmedabad), personal interview with David Kinley, Inst. for Food and Develop. Policy (San Francisco), March 25, 1980.
15. "Report on Plant Protection in Major Food Crops in Malaysia," FAO, Dec. 1977, p. 17.
16. "Basic Supply and Marketing Data for Agro-Pesticides in Indonesia," ARSAP/Pesticides, FAO (Bangkok), Jan. 1980, p. 25.
17. Confidential interview with authors.
18. *Agriculture: Toward 2000*, U.N. Food and Agriculture Organization, Rome, July 1979, p. 82.
19. Eleanor Randolph, "Seed Patents: Fears Sprout at Grass Roots," *Los Angeles Times*, June 2, 1980, p. 1.
20. H.R. 999, introduced Jan. 18, 1979.
21. P.R. Mooney, *Seeds of the Earth* (London: International Coalition for Develop. Action, 1979), p. 57.
22. Ibid., p. 62.
23. Ibid., p. 61.

24. Ibid., pp. 55, 58.
25. Ibid., p. 56.
26. Ibid.
27. Confidential interview with authors.
28. Randolph, op. cit.
29. Ibid. See also *Conservation of Germplasm Resources: An Imperative* Washington: Committee on Germ Plasm Resources, National Academy of Sciences, 1978).
30. Confidential interview with authors.
31. Frances Moore Lappé and Joseph Collins, *Food First: Beyond the Myth of Scarcity* (New York: Ballantine, 1979), p. 157.
32. Ibid.
33. Mooney, op. cit., p. 14.

CHAPTER SEVEN

1. "Banking by Motorcycle in Pakistan," *Farm Chemicals*, Sept. 1979.
2. Dr. M.A.A. Beg, chief scientific officer, Pakistan Council of Scientific and Industrial Research Laboratories, Karachi, letter to authors, Feb. 16, 1980.
3. Confidential interview with authors, March 1980.
4. "O Que Fazer para Acabar com a Prago dos Inseticidas," Sao Paulo *Jornal da Tarde*, Sept. 8, 1975, p. 17.
5. Dr. Fred Whittemore, U.S. AID, personal interview with authors, March 17, 1980.
6. Ibid.
7. *Annual Report 1978*, Interamerican Institute for Agricultural Sciences (IICA), p. 78.
8. Michael Moran, personal interview with authors, March 12, 1980.
9. Ibid.
10. Cited in Robert Stein and Brian Johnson, *Banking on the Biosphere? Environmental Procedures and Practices of Nine Multilateral Development Agencies* (Lexington, Mass.: Lexington Books, D.C. Heath and Company, 1979). This book is a report of a study by the International Institute for Environment and Development conducted with the United Nations Environment Program (UNEP).
11. "Declaration of Environmental Policies and Procedures Relating to Economic Development," adopted in New York on Feb. 1, 1980.
12. Richard Levine, "Growth of World Bank under McNamara Rule Generates Controversy," *Wall Street Journal*, Oct. 5, 1976.
13. Ibid.
14. Confidential interview with authors.
15. Ibid.
16. *Environmental Defense Fund* v. *U.S. Agency for International Development*. "Stipulation," U.S. District Court, District of Columbia, Civil Action No. 75-0500.
17. Cynthia Arnson and William Goodfellow, "OPIC: Insuring the Status Quo," *International Policy Report*, Vol. 3, No. 2 (Washington, D.C.: Center for International Policy, Sept. 1977), p. 2.

18. Frances Moore Lappé and Joseph Collins, *Food First: Beyond the Myth of Scarcity* (New York: Ballantine, 1979), p. 353.
19. Arnson and Goodfellow, op. cit.
20. Speech by California farmer, "Politics of Pesticides Conference," sponsored by the Coordinating Committee on Pesticides, March 29, 1980.
21. "Report on Environmental Assessment of Pesticide Regulatory Programs, CDFA, vol. 1, summary, 1979.
22. Robert van den Bosch, *The Pesticide Conspiracy* (Garden City, N.Y.: Doubleday, 1978), p. 27.
23. Ibid.
24. Ibid. Also, see CDFA report.
25. Frances Moore Lappé and Joseph Collins, *Food First: Beyond the Myth of Scarcity* (New York: Ballantine, 1979), p. 74.
26. "Information on Industry Cooperative Programs," FAO document DDI:G/75/40, March 1975.
27. Lappé and Collins, op. cit., p. 75.
28. Confidential interview with authors.
29. Lappé and Collins, op. cit., p. 74.
30. Ibid.
31. B.B. Watts, "Value of Provisional Registration to Facilitate a Full Evaluation of New Pesticides," *FAO Plant Protection Bulletin*, No. 3, 1978.
32. Don Kimmel, personal interview with authors, March 14, 1980.
33. Walter Simons, telephone interview with authors, March 19, 1980.
34. Kimmel interview.
35. Ibid.
36. "The Pesticides in Common Use in Grain Storage," Report of the FAO Global Survey of Pesticide Susceptibility of Stored Grain Pests, 1976.
37. Confidential interview with authors.

CHAPTER EIGHT

1. Edwin Johnson, EPA deputy assistant administrator for pesticide programs, statement before the subcomm. on foreign agriculture policy, Senate Comm. on Agriculture, Nutrition, and Forestry, May 25, 1978.
2. John Walsh, "EPA and Toxic Substances Law: Dealing with Uncertainty," *Science*, Nov. 10, 1978, p. 598.
3. Christopher Arntzen, EPA Recruiter, telephone interview with authors, Sept. 18, 1980.
4. "Notices of Judgment under the Federal Insecticide, Fungicide, and Rodenticide Act," Nos. 1901-1950, issued by EPA Office of Enforcement, Pesticides and Toxic Substances Division, Nov. 1976.
5. Ibid; Nos. 2001-2050, issued May 1977.
6. 7 U.S.C. §136.
7. "Better Regulation of Pesticide Exports and Pesticide Residues in Imported Food Is Essential," U.S. GAO, Rept. No. CED-79-43, June 22, 1979.

8. S. Jacob Scherr, personal interview with authors.
9. Caroline E. Mayer, "Easing of Hazardous Exports Studied," *Washington Post,* September 9, 1981, and *Hazardous Materials Intelligence Report,* September 18, 1981.
10. Statement by Dr. Jack Early, president, National Agricultural Chemicals Association, before the Subcommittee on Department Operations, Research and Foreign Agriculture of the Committee on Agriculture, U.S. House of Representatives, July 16, 1981.
11. Dr. Hal Hubbard, telephone interview with authors, June 1, 1980.
12. Francis Miley, telephone interview with authors, May 15, 1979.
13. Francis Miley, personal interview with authors, June 5, 1979.
14. Fredrick Rarig, statement in "Proceedings of the U.S. Strategy Conference on Pesticide Management."
15. Samuel Gitanga, personal interview with authors, June 8, 1979.
16. Personal interview with Mark Dowie, March 1979.
17. "Background Report on the Executive Order on Federal Policy Regarding the Export of Banned or Significantly Restricted Substances," pp. 30–31.
18. Ibid.
19. "Comments of Public Citizen Health Research Group on Hazardous Substances Export Policy," May 15, 1980.
20. *Washington Post,* January 16, 1981, and *Science,* March 27, 1981.
21. H.R. 2439, introduced in early 1981.

CHAPTER NINE

1. *The Standard,* quoted in newspaper reports, Nairobi, Kenya, May 11, 1977, p. 3, and April 25, 1979, p. 3.
2. Farmers Assistance Board, *The Proliferation of Dangerous and Poisonous Pesticides in Rice Production (A Guideline for Case Studies),* 1979, p. 1.
3. "Report on Plant Protection in Major Food Crops in Malaysia," FAO, Dec. 1977, p. 2.
4. Personal interview with authors, March 20, 1980.
5. OECD—News release from U.S. Consumer Products Safety Commission, release date June 12, 1980.

FOR MORE INFORMATION

Pesticides and Pills: For Export Only. This hard-hitting two-part documentary probes the dumping of pesticides and medicines in the third world. Made by Robert Richter for PBS, the 1981 film includes an interview with *Circle of Poison* co-author David Weir. The two one-hour segments are available on 16mm film or video. Contact Robert Richter Productions, 330 West 42nd St., New York, NY 10036, (212) 947-1396.

The Cost of Cotton. By using large amounts of DDT and other pesticides, Guatemala has quietly become the world's most efficient producer of cotton. But the costs have been high: numerous deaths and thousands of poisonings result each year from the intensity of the spraying. DDT levels in mother's milk are the highest ever recorded. This probing, 30-minute documentary chronicles the grave health and environmental problems created by this economic miracle. Available in 16mm film or ¾-inch video tape; French and Spanish versions also available. Winner, Lille International Film Festival; Finalist, American Film Festival. Available from Jim Watson, 176 Linda St., San Francisco, CA 94110, (415) 285-5979, or Luis Argueta, 444 West 46th St., Apt. 1-D, New York, NY 10036, (212) 245-7538.

Pills, Pesticides and Profits: The International Trade in Toxic Substances by Karim Ahmed, S. Jacob Scherr and Robert Richter. A major book on international dumping, to be published in early 1982 by North River Press, Box 241, Croton-on-Hudson, NY 10520, $10.95.

"Export of Hazardous Products," Hearings before the Subcommittee on International Economic Policy and Trade, House Committee on Foreign Affairs, 96th Congress, 2nd session, June 5 and 12, Sept. 9, 1980.

Information packet on dumping available from the Natural Resources Defense Council, 1725 I St. NW, Suite 600, Washington DC 20006, (202) 223-8210.

INSTITUTE PUBLICATIONS

Now We Can Speak: A Journey through the New Nicaragua, features interviews with Nicaraguans from every walk of life telling how their lives have changed since the 1979 overthrow of the Somoza dictatorship. Frances Moore Lappé and Joseph Collins, 124 pages with photographs.
$4.95

What Difference Could a Revolution Make? Food and Farming in the New Nicaragua, provides a critical yet sympathetic look at the agrarian reform in Nicaragua since the 1979 revolution and analyzes the new government's successes, problems, and prospects. Joseph Collins and Frances Moore Lappé, with Nick Allen, 185 pages.
$5.95

Trading the Future: Farm Exports and the Concentration of Economic Power in Our Food System is a scholarly investigation which develops a comprehensive analysis of U.S. farming and food systems. It demonstrates how the increasing concentration of control over farmland, rapid erosion of soil, loss of water resources, and our growing reliance upon a narrow range of export crops parallels the process of underdevelopment experienced in the third world. Alterations in America's farm landscape threaten us not only with severe imbalances in control over resources, but also with rising prices in the midst of huge surpluses. James Wessel with Mort Hantman, 250 pages.
$8.95.

Diet for a Small Planet: Tenth Anniversary Edition, an updated edition of the bestseller that taught Americans the social and personal significance of a new way of eating. Frances Moore Lappé, 432 pages with charts, tables, resource guide, recipes, Ballantine Books.
$3.50

Food First: Beyond the Myth of Scarcity, 50 questions and responses about the causes and proposed remedies for world hunger. Frances Moore Lappé and Joseph Collins, with Cary Fowler, 620 pages, Ballantine Books, revised 1979.
$3.95

Comer es Primero: Mas Alla del Mito de la Escasez is a Spanish-language edition of *Food First*, 409 pages, Siglo XXI—Mexico. $9.95

Food First Comic, a comic for young people based on the book *Food First: Beyond the Myth of Scarcity*. Leonard Rifas, 24 pages. $1.00

Aid as Obstacle: Twenty Questions about our Foreign Aid and the Hungry demonstrates that foreign aid may be hurting the very people we want to help and explains why foreign aid programs fail. Frances Moore Lappé, Joseph Collins, David Kinley, 192 pages with photographs. $5.95

Development Debacle: The World Bank in the Philippines, uses the World Bank's own secret documents to show how its ambitious development plans actually hurt the very people they were supposed to aid—the poor majority. Walden Bello, David Kinley, and Elaine Elinson, 270 pages with bibliography and tables. $6.95

Against the Grain: The Dilemma of Project Food Aid is an in-depth critique which draws extensively from field research to document the damaging social and economic impacts of food aid programs throughout the world. Tony Jackson, 132 pages, Oxfam—England. $9.95

World Hunger: Ten Myths clears the way for each of us to work in appropriate ways to end needless hunger. Frances Moore Lappé and Joseph Collins, revised and updated, 72 pages with photographs. $2.95

El Hambre en el Mundo: Diez Mitos, a Spanish-language version of *World Hunger: Ten Myths* plus additional information about food and agriculture policies in Mexico, 72 pages. $1.45

Needless Hunger: Voices from a Bangladesh Village exposes the often brutal political and economic roots of needless hunger. Betsy Hartmann and James Boyce, 72 pages with photographs. $3.50

Circle of Poison: Pesticides and People in a Hungry World documents a scandal of global proportions, the export of dangerous pesticides to Third World countries. David Weir and Mark Schapiro, 101 pages with photos and tables. $3.95

Circulo de Veneno: Los Plaguicidas y el Hombre en un Mundo Hambriento is a Spanish-language version of *Circle of Poison*, 135 pages, Terra Nova—Mexico. $3.95

Seeds of the Earth: A Private or Public Resource? examines the rapid erosion of the earth's gene pool of seed varieties and the control of the seed industry by multinational corporations. Pat Roy Mooney, 126 pages with tables and corporate profiles. $7.00

A Growing Problem: Pesticides and the Third World Poor, a startling survey of pesticide use based on field work in the Third World and library research. This comprehensive analysis also assesses alternative pest control systems. David Bull, 192 pages with charts, photos, and references. $9.95

What Can We Do? An action guide on food, land and hunger issues. Interviews with over one dozen North Americans involved in many aspects of these issues. William Valentine and Frances Moore Lappé, 60 pages with photographs. $2.95

Mozambique and Tanzania: Asking the Big Questions looks at the questions which face people working to build economic and political systems based on equity, participation, and cooperation. Frances Moore Lappé and Adele Negro Beccar-Varela, 126 pages with photographs. $4.75

Casting New Molds: First Steps towards Worker Control in a Mozambique Steel Factory, a personal account of the day-to-day struggle of Mozambique workers by Peter Sketchley, with Frances Moore Lappé, 64 pages. $3.95

Agrarian Reform and Counter-Reform in Chile, a first-hand look at some of the current economic policies in Chile and their effect on the rural majority. Joseph Collins, 24 pages with photographs. $1.45

Research Reports. "Land Reform: Is It the Answer? A Venezuelan Peasant Speaks." Frances Moore Lappé and Hannes Lorenzen, 17 pages.   *$1.50*

"Export Agriculture: An Energy Drain." Mort Hantman, 50 pages.   *$3.00*

"Breaking the Circle of Poison: The IPM Revolution in Nicaragua." Sean L. Swezey and Rainer Daxl, 23 pages. *$4.00*

Food First Curriculum Sampler offers a week's worth of creative activities to bring the basics about world hunger and our food system to grades four through six. 12 pages. *$1.00*

Food First Slideshow/Filmstrip in a visually positive and powerful portrayal demonstrates that the cause of hunger is not scarcity but the increasing concentration of control over food producing resources, 30 minutes.

$89 (slideshow), *$34* (filmstrip)

Write for information on bulk discounts.

All publications orders must be prepaid.

Please include shipping charges: 15% of order for U.S. book rate or foreign surface mail, $1.00 minimum. California residents add sales tax.

Food First Books

Institute for Food and Development Policy
1885 Mission Street
San Francisco, CA 94103 USA
(415) 864–8555

ABOUT THE INSTITUTE

The Institute for Food and Development Policy, publisher of this book, is a nonprofit research and education center. The Institute works to identify the root causes of hunger and food problems in the United States and around the world and to educate the public as well as policymakers about these problems.

The world has never produced so much food as it does today—more than enough to feed every child, woman, and man as many calories as the average American eats. Yet hunger is on the rise, with more than one billion people around the world going without enough to eat.

Institute research has demonstrated that the hunger and poverty in which millions seem condemned to live is not inevitable. Our Food First publications reveal how scarcity and overpopulation, long believed to be the causes of hunger, are instead symptoms—symptoms of an ever-increasing concentration of control over food-producing resources in the hands of a few, depriving so many people of the power to feed themselves.

In 55 countries and 20 languages, Food First materials and investigations are freeing people from the grip of despair, laying the groundwork—in ideas and action—for a more democratically controlled food system that will meet the needs of all.

An Invitation to Join Us

Private contributions and membership dues form the financial base of the Institute for Food and Development Policy. Because the Institute is not tied to any government, corporation, or university, it can speak with a strong independent voice, free of ideological formulas. The success of the Institute's programs depends not only on its dedicated volunteers and staff, but on financial activists as well. All our efforts toward ending hunger are made possible by membership dues or gifts from individuals, small foundations, and religious organizations. We accept no government or corporate funding.

Each new and continuing member strengthens our effort to change a hungry world. We'd like to invite you to join in this effort. As a member of the Institute you will receive a 25 percent discount on all Food First books. You will also receive our triannual publication, *Food First News*, and our timely Action Alerts. These Alerts provide information and suggestions for action on current food and hunger crises in the United States and around the world.

All contributions to the Institute are tax deductible.

To join us in putting Food First, just clip and return the attached form to the Institute for Food and Development Policy, 1885 Mission Street, San Francisco, CA 94103, USA.

☐ Yes, I want to ensure that the Institute for Food and Development Policy continues to be an independent and effective voice in the struggle against hunger and food problems. I have enclosed my tax-deductible contribution of:

☐ $20   ☐ $35   ☐ $50   ☐ Other $_____

☐ Please send me more information about the Institute, including your publications catalog.

Name _____

Address _____

City _____ State _____ Zip _____ Country _____

Institute for Food and Development Policy
1885 Mission Street
San Francisco, CA 94103 USA

NATIONAL UNIVERSITY LIBRARY

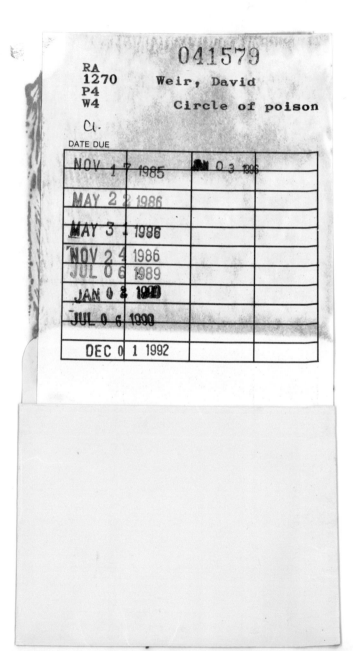